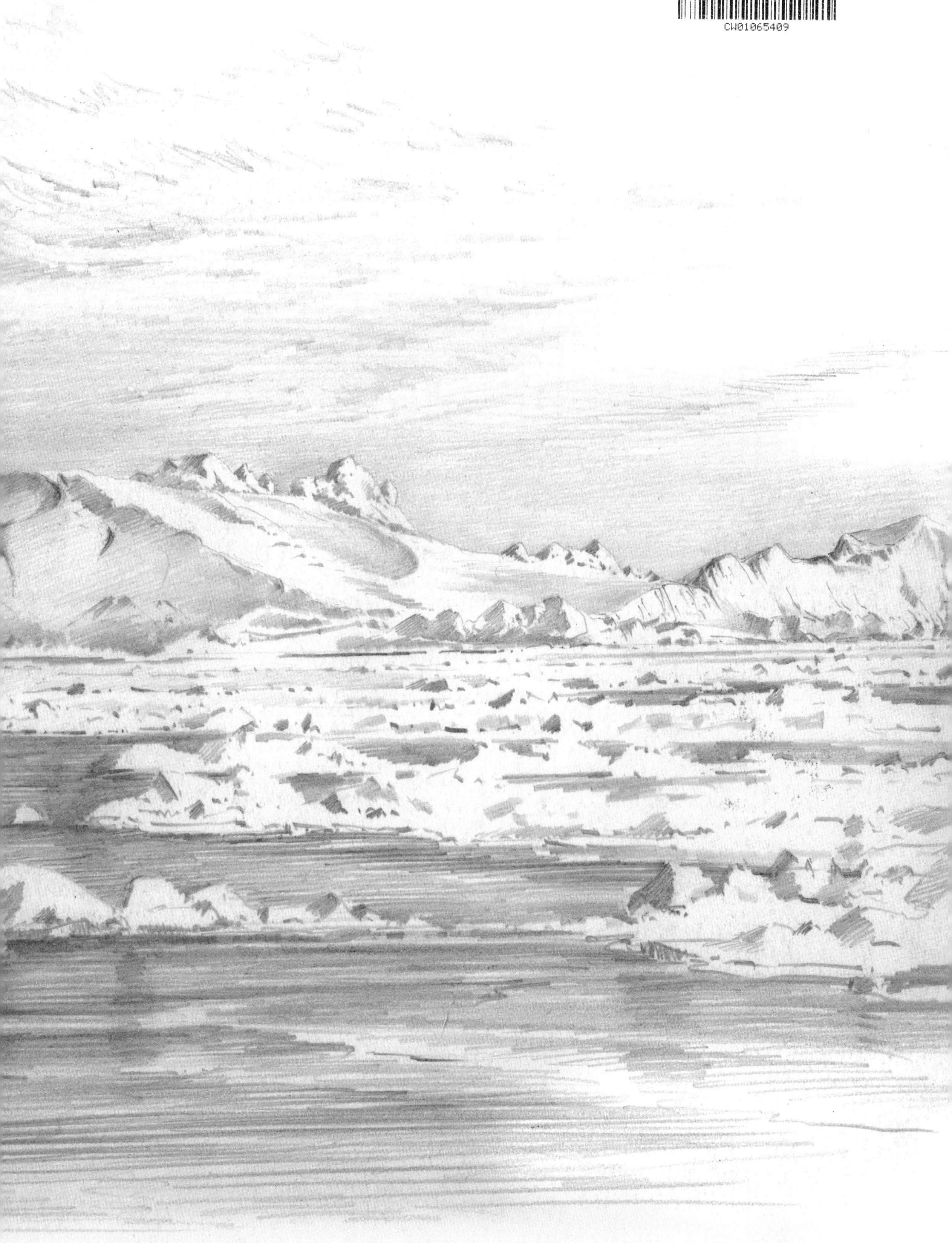

THE BLACK & WHITE SKETCH BOOK
Winsor and Newton Ltd. LONDON.
MADE IN THE FOLLOWING SIZES
3½ × 3
5 × 3¾ | 8 × 5
7¼ × 4½ | 10 × 8

"These days are with one for all time - they are never to be forgotten - and they are to be found nowhere else in all the world but at the poles. The peace of God which passes all understanding reigns here in these days. One only wishes one could bring a glimpse of it away with one with all its unimaginable beauty." EAW

Edward A Wilson

SIZE OF THIS SKETCH BOOK, 7¼ × 4½ in.
DIMENSIONS DE L'ALBUM
GRÖSSE DIESES SKIZZENBUCHES } 18.4 × 11 CM.

Published By
REARDON PUBLISHING
PO Box 919, Cheltenham GL50 9AN England
www.reardon.co.uk

Researched and Edited by
D.M.Wilson and C.J.Wilson

Text by D.M. Wilson © 2000: 2004: 2011
The moral right of the authors has been asserted.

Copyright © 2011

Layout and Design by Nicholas Reardon

ISBN 9781874192510 Hardback Edition
ISBN 9781874192572 Special Limited Edition

Printed through
World Print Ltd
Hong Kong

Antarctic (Discovery) Exp.   Birds, Plate IX.   E.A.Wilson pinx.
Bale & Danielsson. Ltd lith.

EDWARD WILSON'S ANTARCTIC NOTEBOOKS

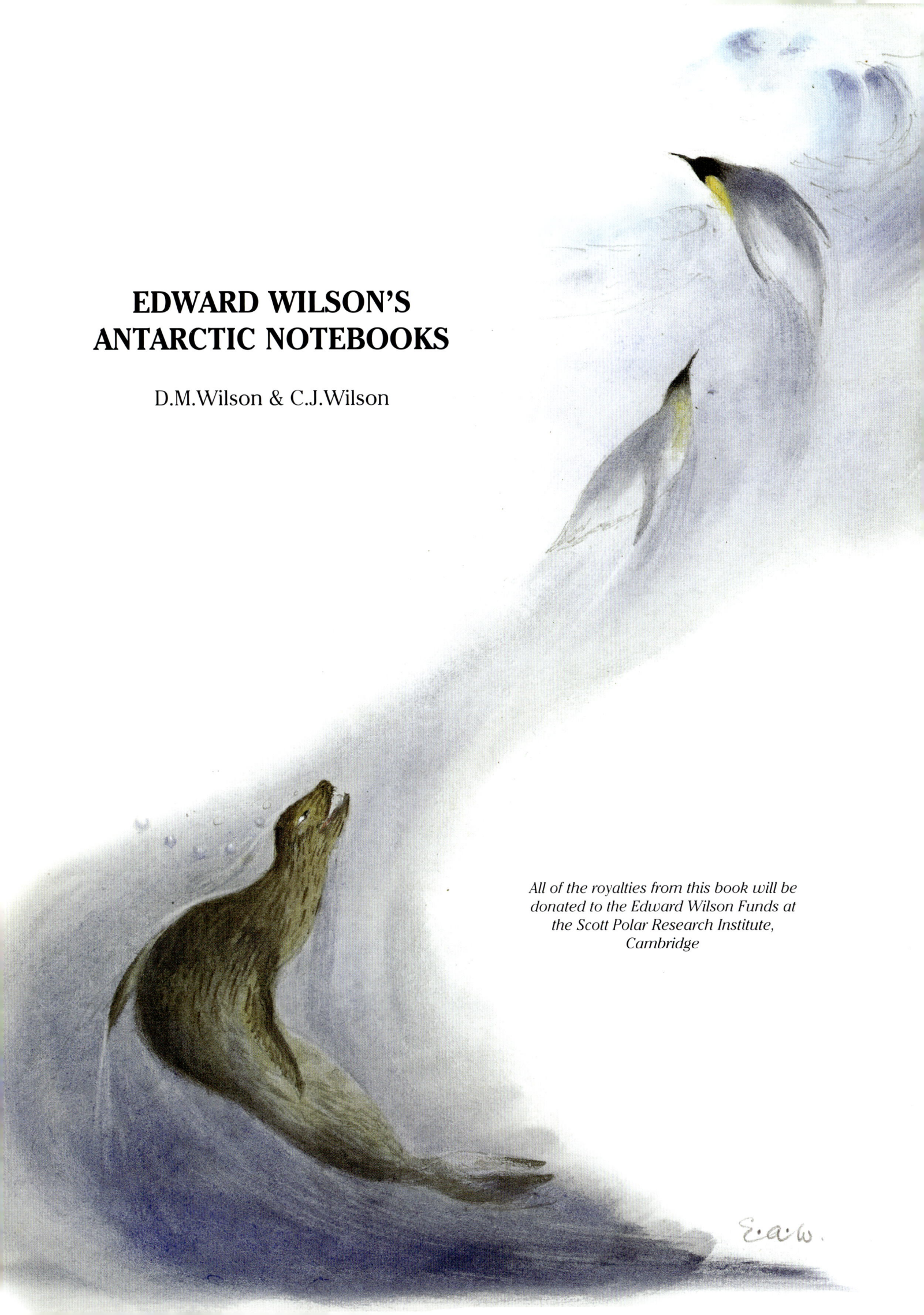

D.M.Wilson & C.J.Wilson

All of the royalties from this book will be donated to the Edward Wilson Funds at the Scott Polar Research Institute, Cambridge

Caught in a Blizzard

Introduction

Dr. Edward Adrian Wilson (1872-1912) opened up a continent with his pencils and paint-box. He recorded in detail the events and discoveries with which he was intimately involved at the height of the Heroic Age of Antarctic Exploration, being aboard both of Captain Scott's Antarctic Expeditions, *Discovery* (1901-1904) and *Terra Nova* (1910-1913). Appointed for being a medical doctor and naturalist, he became one of the last major practitioners of exploration art. This artistic form explored the unknown world and illustrated it for the public back home, allowing European horizons to expand and defining the template of their visualisation. Its mode was coalesced by the fusion of scientific, cartographic and artistic techniques by the artist William Hodges, during Captain Cook's 2nd Expedition (1772-1775). Its practitioners delivered capable artworks in all these fields, accurately recording everything discovered on expeditions: landscapes, topographical coastlines, scientific specimens, events, portraits of expedition life etc. as accurately as possible. Expedition artwork peaked during the nineteenth century, at the height of Imperial expansion but it is doubtful whether any of its practitioners realised that they were part of what is now an identifiable artistic movement, least of all Edward Wilson who never believed himself to be more than an amateur dabbler. He admired and studied in great detail the work of William Turner, whom he considered the greatest of landscape painters, he was also inspired by Japanese water-colourists and was a disciple of John Ruskin; nevertheless, Edward Wilson was a self-taught artist who felt the lack of training greatly deprecated his own talents. With his death alongside Captain Scott during the return from the South Pole in 1912, the heart of the last of the great continents had been revealed, whilst with the major media for recording feats of exploration passing primarily to photograph and film, exploration art became submerged into the realm of the aesthetic.

Dr. Edward Adrian Wilson (1872-1912) is also our great-uncle. All our lives he has been to us simply, Uncle Ted. We grew up with his pictures on our walls and artefacts from his life and expeditions in our home. This seemed to us perfectly normal, he was our grandfather's older brother and that was that. Times change, however, and now many of these treasures are in Museums for safekeeping and so that they may be enjoyed by all. Uncle Ted liked his pictures to be seen. Indeed, he insisted on it. He thought his modest talent a gift from God, to be shared. As a result, he also hated them being sold, although he sold them when he had to. Generally, however, his pictures were given away, a practise kept up by many in the Wilson family even today, out of respect for his wishes. His desire for his pictures to be seen had some interesting repercussions. He gave almost none to the Royal Geographical Society, despite their patronage of Scott's expeditions, simply because he said that they would be 'hidden in a cupboard under the stairs'. It also meant that when his widow, Aunt Ory, was persuaded to give many of her husband's watercolours to the newly founded Scott Polar Research Institute in Cambridge in the 1920's it was on the basis that they were to be displayed. A whole gallery floor was built for the purpose and the Ruskin cabinets, which had been used to show Turner's pictures in the Tate Gallery, were given for the purpose of exhibiting Wilson's watercolours. The exhibit was hugely popular for many years. However, Uncle Ted painted in watercolour, sometimes highlighted with gouache. He also drew in pencil, charcoal, chalk and occasionally pastel. His works were almost always executed on paper, occasionally card. As far as we can ascertain, he never used oils nor board. Watercolour, however, is damaged very quickly by exposure to the light, as is paper. For their own safekeeping, therefore, Uncle Ted's pictures were generally removed from public display. They are almost all now in 'cupboards under the stairs', all be they archival quality cupboards. This has set us with something of a problem: how may his pictures be seen and enjoyed, as he wished? One obvious answer is publication.

Many of his pictures have been published over the years but to all who know the original images, no publication does them justice. The subtlety of his work and colouration is almost impossible to replicate. Technology, however, is starting to overcome this hurdle. We first produced a volume of his non-Antarctic pictures, *Edward Wilson's Nature Notebooks*, in 2004, reproducing a wide selection of his non-Antarctic work. We reasoned that his Antarctic work was quite well known. We were delighted that digital photography allowed us to edit and re-organise the pictures from his notebooks into chronological order rather than the scientific order in which he kept them, allowing us to produce the first biography of Uncle Ted through his pictures. This proved to be a very popular volume but almost everybody asked us when the Antarctic pictures were going to follow. So here they are. The format closely replicates the *Nature Notebooks*, as the *Antarctic Notebooks* essentially represent the two missing chapters from that volume. Likewise, in this volume, the chapters on his non-polar work are minimal, as they are given substantial treatment in the *Nature Notebooks*. Whilst each has been designed so that it can stand alone, the two volumes are really designed to go together. Despite having given a single volume to the Antarctic pictures and additional pages, even now, we are very conscious of all the images that we have left out. This volume only gives an representative selection of what he accomplished but it does draw together a large number of his Antarctic works, from both expeditions, into one volume for the first time. Indeed, a number of the images receive their first publication here.

It sounds simple but, of course, it is a huge undertaking. The sheer number of individuals and institutions that hold his works are considerable. More daunting however, is the fact that digital photography can only go so far. In line with his wishes, his pictures have very often been hung and large numbers of them are severely damaged. Our father (J. Michael Wilson) always used to say that it is the pinks that go first - and he is right. The pinks fade from Uncle Ted's paintings almost so fast that you can watch them go; they are followed by other colours until the pictures become drab and lifeless. When you do come across a painting that has not been hung it is breathtakingly vibrant, with delicious subtleties. It is not just light that damages the colour, often the paper has been greatly acidified by backing boards and frames, perverting the colours. Foxing, scratches, mould, dirt and other damage are also frequent issues. In some instances older prints of the paintings are better representations of his work than the original itself, such is the corruption. So what to do, when a digital photograph is actually a perfect representation of a degraded picture? We feel sure, that had Uncle Ted been alive, he would have repainted the images for publication. Such was his fine sense of colour, akin to a musician with perfect pitch, that he had already noted his intention to repaint every single painting from the *Terra Nova* expedition, as many had been executed under an acetylene lamp with a yellow tinge to the light, which distorted his colours. Some of the images are clearly marked as such. However, he is not alive to do it, so it has fallen to us to try to reproduce his pictures in a form which we feel he would have approved of, a form that is fresh and alive, not acidified and light degraded. Fortunately, digital technology has helped us, once again. Where necessary, we have removed blemishes and restored the colouration. Sometimes the colour still lurks in a small part of the picture hidden under a frame from the light, which allows restoration to the rest of the image; sometimes it is residual in the image; sometimes it is preserved in period prints; and often it can be related to his original pencil sketch with his colour notes, which can guide restoration. Occasionally, there is no information from any of these devices in which case it cannot be restored with anything like a nod to the original image. These we have left - or left out of the book all together. So the reader should be aware that actually many of the reproductions in this book are superior to the pictures in the archives. We have gone to great efforts to maintain accuracy and hope that we have achieved this, however, for technical reasons, there are bound to be discrepancies for which we can but apologise.

Partly as a result of our wishing to reproduce the scrapbook feel of the original notebooks - with pictures leaping out from every corner in an almost bewildering array, and also because of technical limitations, we have not attempted to reproduce these images to scale. Additionally, many of Uncle Ted's notebooks are widely annotated, so we have reproduced a selection of quotations that give a reflection of his thinking and natural history observations. Additionally, there is a short biographical text at the start of each chapter based on the text of the *Nature Notebooks*. We hope that this will interest many to engage in further reading and exploration of our great uncle's writings and paintings.

Nevertheless, as with its companion volume, ultimately the *Antarctic Notebooks* is peculiarly personal. It isn't particularly scientific, nor very 'arty' and it is unashamedly non-academic. For this we make no apology - we simply wanted to share the privilege that we grew up with and to make it possible for other people to enjoy a wider range of Uncle Ted's pictures.

We hope very much that you will enjoy sharing them with us.

D.M.Wilson and C.J.Wilson

Please open the door at once! He's thawed!

Acknowledgements

The pictures of Edward Adrian Wilson are widely dispersed. This is, in large part, due to the generosity of members of the Wilson family who, over the course of a hundred years, have donated his pictures to many of our public collections and given them freely to friends. However, it means that in order to produce a representation of his Antarctic notebooks today, the co-operation of a large number of institutions and private individuals is required.

In particular, we are deeply honoured to include pictures from the collection of Her Majesty, Queen Elizabeth II. We are grateful to Her Majesty and also to Karen Lawson and Emma Stuart of the Royal Collections for their assistance. We are further indebted to the Director and staff of the Scott Polar Research Institute, University of Cambridge, in particular to the Archivist, Naomi Boneham who has provided extensive support; to Lucy Martin, Don Manning and to Heather Lane. We are grateful to the staff of the Cheltenham Art Gallery and Museum, in particular to Helen Brown and Ann-Rachel Harwood; to the staff at the Natural History Museum in London, principally Alice Kirk; Douglas Russell; and Eloise Donnelly; and to the Royal Geographical Society (with IBG), notably Jamie Owen; to the Headmaster of Cheltenham College and the College Archives, especially to Christine Leighton; to Chris Jones, Headmaster of Copthorne School and to Debra Okitikpi, Head teacher of the Edward Wilson Primary School, London. The Bridgeman Art Library has an extensive array of Wilson images, including pictures previously sold at Christie's, we gratefully acknowledge the use of these. We are also obliged for the support of Nick Rogers and the Abbot Hall Art Gallery in Kendall; of Linda Noble and the National Marine Biological Library, Plymouth; and of Gill Poulter of the Dundee Heritage Trust. We also deeply appreciate being able to reproduce pictures from the Andrew Croft Memorial Fund collections, with particular thanks to Julia Korner. We have also used photographs which are copyright to private collections but for which the original pictures have since been donated to the Bushey Museum, Herts. and to the Canterbury Museum, Christchurch, New Zealand.

We are grateful to all the above for permission to reproduce images from their collections, for their often generous support and for the assistance and advice of their staff.

We warmly thank, our publisher, Nicholas Reardon, for the considerable effort that he has undertaken to make this project a reality.

Finally, we would like to thank the following for assistance with photographic work, for permissions to reproduce images, for their support, advice and/or endless patience: Paul Davies of Kingsbridge Books, Dr. Desna Greenhow; Phillip Scott; John Scott; Judy Skelton; Judy Barker; Muriel Finnis; Tom Wilson; Mark Wilson; - and last, but hardly least, for the love and support of Duncan Lawie FRSA and Ann M. Wilson ('our Annie').

Needless to say, this book is our own, and responsibility for any errors and editorial choices rests with us.

Editors' Notes

In keeping with the historic period, all units of measurement are given in imperial values with the metric conversion following in brackets. Please note that during the Heroic Age of Antarctic exploration geographical (or nautical) miles were generally used. One geographical mile is equivalent to 1.15 statute miles or 1.85 kilometres.

Stanmore Common

The Formative Years: 1872-1901

"The more we try the clearer becomes our insight, and the more we use our thinking faculties the quicker they become in their power of grasping points of Truth." EAW

The Formative Years: 1872-1901

Edward Adrian Wilson was born on 23 July 1872 at Montpellier Terrace, Cheltenham. He was the second son and fifth child of Edward Thomas Wilson (1832-1918) and his wife, Mary Agnes, née Whishaw (1841-1930). In the family he was simply known at Ted.

On his mother's side, the Whishaw family constituted a long line of successful lawyers and businessmen with particular ties to Imperial Russia. On his father's side, the Wilson family were wealthy Trans-Atlantic Quaker industrialists and philanthropists with natural history interests. They founded railways and banks and endowed the Philadelphia Academy of Natural Sciences. A series of bad investments meant that Ted's father did not inherit a fortune but was merely comfortable. Edward Thomas Wilson settled as a pioneering physician in Cheltenham. Ted's childhood was full of news from his Wilson and Whishaw relatives, from the far flung corners of Imperial Russia, the British Empire and beyond. Uncle Charlie (Major General Sir Charles Wilson 1836-1905) was a particularly good source of stories: he served with the Royal Engineers and was sent to relieve Gordon at Khartoum.

Edward Wilson, c 1873

From the age of 3 his parents noted that Ted liked nothing better than to lie on the floor drawing. This resulted in his mother giving him drawing lessons. From the age of five onwards it is clear that the results were thought to be of interest, as they were collected into scrapbooks. The family often took long country walks and with his father as his guide, the Gloucestershire countryside became Ted's first love and inspiration. By the age of 9 he had announced that he was going to become a naturalist and his mother noted that he would far rather have a "naturalists' walk" with his father than play the games of the playground, despite the fact that he was good at them. Ted started to develop his own natural history collections, collecting "everything he can lay his hands on". He received his first lessons in taxidermy at the age of 11.

Becoming a day pupil at the Cheltenham Proprietary College for Boys, his academic work was solid but never extraordinary. He excelled at games, at art and in the activities of the Natural History Society, being secretary of the ornithological section for some time. During this period, his mother, a noted breeder of poultry, took on the lease to a large farm, *The Crippetts*, near Shurdington. The fact that he was a day pupil allowed him to wander freely before and after school in its fields and hedgerows, making observations of the natural world, its wildlife and the changing seasons. He would often wrap himself in a cloak and leave the house before dawn to be in position in the woods before sunrise or take his supper up to *The Crippetts* to observe rabbits playing in the long summer evenings. Through his teenage years he taught himself to become a quite remarkable field naturalist. It is said that he not only learned to recognise the calls of the birds, but could state exactly what the bird was doing when it made such a call. Despite his obvious interests, it was decided that Ted would enter Gonville and Caius College in Cambridge to pursue a career in medicine, like his father. Natural History and art were clearly seen as subjects for hobbies and not a career. In some ways Ted struggled with this belief for the rest of his life.

The intake of freshmen to Gonville and Caius College in 1891 quickly developed a reputation for being turbulent. Ted's sense of humour won him many friends, as did his generosity with his knowledge and possessions. Ted also quickly developed a reputation amongst his tutors and peers as a mediator and peacemaker, a role that was to follow him for the rest of his life: his fellows trusted his moral judgement and his personal integrity. Ted knuckled down to his studies, the several hours that he devoted to study every day soon paying off as he was awarded the status of an Exhibitioner. Ted also rowed for his College, scoring several pewters. His contemporaries commented that his rooms looked like a Museum they were so full of interesting skulls, bones, feathers, plants and other specimens. The walls and floors were littered with pencil and chalk drawings, which were his preferred media at this time. Nevertheless, he drew in pen and ink and was increasingly experimenting with watercolour. He amused visitors to his rooms by producing their portraits in watercolour, pencil or silhouette. Holidays were spent either at *The Crippetts* or travelling in Europe, devouring the art and architecture of Rotterdam, Antwerp, Brussels, Cologne and Cassel.

It was during this time that Ted began to formulate the deep moral code by which he lived his life. This was based upon a stringently ascetic reading of the New Testament. None but his intimate friends ever realised that it was this deeper spiritual level that gave Ted the moral power they so respected, as he rarely talked of such things, yet it gave him the inner strength and ascetic self-discipline upon which his future life would depend. Ted came to care little for originality and greatly for Truth, whether scientific, moral, artistic, spiritual or physical. Every aspect of life became, for him, a part of an indivisible Divine Truth with science, art and poetry simply different ways of explaining and experiencing a complex but Divine creation.

In 1894 Ted sat his exams, taking his degree with a first class pass in the Natural Science Tripos (part I). He was, quite naturally, utterly delighted and chose 5 volumes of Ruskin as a prize: his reading of Ruskin was becoming increasingly influential upon his own views on art. Ted was expecting to go straight on to St. George's Hospital to work towards becoming an FRCS and was astonished when the Master of Caius asked his parents if he might be permitted to stay on at Cambridge for another year because he was considered to be such a good influence on the College. Ted was not in favour but he was persuaded to stay on at Cambridge for an extra year. Perhaps this was due in part to the sudden death of his youngest sister, Gwladys and Ted not wishing to add further to his parents concerns.

Edward Wilson, c 1882

Ted pursued his love for art and the natural world throughout his remaining time at Cambridge, rather than academic success. He increasingly indulged his passion for walking in the fens, sketching the wildlife and making notes. His reading became pre-occupied with biographies of the great artists in addition to his medical books. A consequence of this was the life-changing realisation that even the greatest artists had had to study and learn their artistic technique. This realisation was liberating for Ted and began what his father suggested was a 'craze' for drawing and art. He started to work and to re-work his pictures, training himself in artistic techniques by seeing and criticising the work of others and accepting the critique of his family and friends. A new desire, however, was for art lessons that would train him to be a 'proper' artist and he harboured plans to take evening classes at an art school during his hospital training in London.

Working for the second part of his MB degree, involved long hours of hard work at St. George's Hospital and he was soon immersed in anatomy, physiology and surgery. Nevertheless, he found time to play football for the hospital and also to row. He took lodgings in Paddington and walked to and from St. George's which was at Hyde Park Corner, often taking detours through the London parks or the Zoo to hunt for "weeds and mushrooms" to draw and paint. Yet for all of its stimulation, London felt claustrophobic to him. London, he wrote, particularly in the spring, left him feeling like a "soda-water bottle in an oven". Additionally, Ted must have felt family pressures intensely at this time, with one of his sisters marrying and another dying of Typhoid whilst nursing an epidemic in Leicester.

For Ted, principles had to be lived, not to remain mere ideals. He strove to instantiate Christian principles in his life and this involved, in part, living a frugal existence. He prided himself on living on as little money as possible and when he had it, he often simply gave it away. As a result, his watch, or other jewellery was often temporarily deposited with the pawnbroker. Perhaps this is why he never engaged in his desired evening classes at art school, or perhaps it was simply that he had too much to do. Already, his midwifery course required long hours in the slums and was depriving him of sleep.

Edward Wilson, c 1891

Ted spent what time he could in painting and drawing or in studying the works of the great artists in the London galleries. He admired the work of Japanese water-colourists but thought little of Tissot or the artists of the Paris Salon. Most importantly, however, he became "smitten to distraction" by Turner and spent many hours studying his pictures. Turner was to become an enormous influence on Ted's artistic technique, just as Ruskin had become a huge influence on the way that Ted thought about art. He increasingly sought to express these ideals in his pictures. Truth in painting, the ability to paint or draw whatever is seen as accurately as possible, became the benchmark of his success or failure. If he did not record something well enough, he tore up the result, no matter how aesthetically pleasing the picture might appear. So tightly was drawing or painting tied up with Truth for Ted, that its execution became for him a form of prayer or meditation, with the result that he increasingly hated selling his pictures, preferring to give them away. These deeper motivations and principles on which he based his life, however, were hidden from all but his closest friends. To his colleagues he was his usual jovial self.

Ted's artistic abilities were soon noted at St. George's. His tutors and peers delighted in his predilection for drawing portraits of them during lectures and this soon lead to requests for the use of his talents. In particular, his skills came into demand for pathological drawing, his accurate illustration of diseased organs being utilised by the hospital museum and to illustrate papers by several of his tutors. Dr. Rolleston asked him to illustrate his book on the *Diseases of the Liver*, which he happily agreed to, although this wasn't finally published until 1905. Another colleague asked him to assist with some illustrations for his book on fishing flies.

In the autumn of 1896 he moved his lodgings, taking up residence in the Caius Mission house in Battersea. Here he became engaged in youth clubs and Sunday school classes for the children of the Battersea slums. Ted was also starting to seriously consider the idea of becoming a missionary in Africa. He believed that his skills for mission work, doctoring and natural history sketching might usefully be given "free elbow room" in a life there, whereas he had his doubts about being in a position to combine all of his interests in a career in England. This thought was not well received by his parents and he was eventually persuaded to postpone taking any immediate action upon the matter. It was at the Mission house, one afternoon, that he first met Miss Oriana Souper, a friend of the Warden's wife. Even he did not realise how smitten he was at the time.

When Ted returned to Cheltenham for the Christmas of 1897, his parents thought that he was looking frail and worn. On New Years Day, 1898, he started to read a biography of St. Francis of Assisi who rapidly became a role model for him. He was soon back at his hospital duties in London and also took on most of the work of the Mission, due to the illness of the warden. Given the intensity of his life in London, it seems little wonder that his health began to waver. By February of 1898 he was suffering a high fever, pains and giddiness, but he continued to work. After 3 weeks, he went and consulted with Dr. Rolleston who, having performed various tests, confirmed that he had demonstrated the bacilli; Ted had contracted Pulmonary Tuberculosis. He was advised to seek treatment at Davos at once. It was a considerable shock both to Ted and to his family, Tuberculosis frequently being fatal.

After a week in bed Ted felt better and was considered strong enough to go to *The Crippetts*. Ted spent his time in drawing and painting the countryside that was dearest to his heart, producing the most astonishing studies of Cotswold wildlife. Often he would sit in the woods at dawn, wrapped in his cloak, observing the natural world, with "*Modern Painters*, the New Testament and a good deal of pain" for company. He wrote that the thought that death was within "measurable distance" brought him "extra-ordinary peace of mind". The vibrancy of his painting suggests that Ted was far from morbid however, and that he was starting to make real progress.

Through some family friends, Ted was invited to Norway for the summer. The main thrust of tuberculosis treatment at this time involved plenty of rest and cool, dry air. Dr. Rolleston thought that the air of northern Norway was as likely to be as good as the air of Switzerland and so Ted's trip to the sanatorium at Davos was postponed. He was soon in Norway, just south of the Arctic Circle, staying near to modern day Brønnøysund. Here Ted could do more or less as he pleased and delighted in tramping the vast areas of moorland and forest, collecting specimens and sketching. Ted returned to Cheltenham in August but his tuberculosis was not clear and so in October, Ted was dispatched to the sanatorium at Davos in Switzerland.

The Sanatorium was something of a shock to Ted. His doctors insisted that he should do nothing. He tried to view his enforced idleness as a penance but he kicked against it at every opportunity. Perhaps as a result, he found the first snow somewhat depressing but soon came around to the challenge of trying to sketch the Alpine scenery and the complex colours of snow. During these periods of inactivity he read and meditated upon the life of St. Francis or sketched. It was during this time at Davos that Ted consolidated the principles of ascetic mysticism by which he lived his life. It was here too, that he fully developed a colour memorisation technique which he had first conceived at Cambridge and now came to fruition. It meant that he was able to make pencil sketches with colour notes and accurately paint up the picture later. It took quite a bit of practice and to begin with he often got things wrong but the one thing that Ted now had time for was to practice his artistic techniques. As Spring came to the Alps and the birds and animals returned to the mountains, Ted's strength began to return. He was allowed to wander far and wide and sketched almost continuously. It was with some relief however, that in May he left the medical regime of Davos and headed home, where he stayed for only a short while before heading on a second visit to his friends in Norway.

Pronounced clear of the disease, in early October Ted returned to London to take up his interrupted medical studies. Once again he took rooms in Paddington but his cough returned with a vengeance and he returned to Cheltenham. It did concentrate his mind on one aspect of his life, however, and he and Miss Souper, who had been in regular correspondence, were soon engaged to be married. Both Ted and Ory were delighted with this somewhat unexpected turn of events.

Ted returned to his medical studies once more, this time taking lodgings at Stanmore in order to avoid the pollution of London. He now had two months before he had to re-sit the second part of his MB exam and eighteen months work to catch up on. It is of considerable credit that he managed to pass. He then worked on writing up his thesis, *Yellow Atrophy of the Liver*.

It was whilst at Stanmore that Ted took a long hard look at his Natural History drawings and decided that anyone could paint or draw a dead bird in a classic Victorian-style plumage study. What he really wanted to do was to be able to capture the essence of the live bird on paper. He therefore set out to re-teach himself to paint and draw. At the same time he started to execute illustrations for *The Lancet* and *Land and Water*. Ted also returned to drawing at the Zoo. Here he came to the attention of the Secretary of the Zoological Society, Dr. Sclater who invited Ted to a meeting of the British Ornithological Union. He was subsequently elected as a member. It was at meetings of the Union that he met Thorburn and Lodge, who admired and criticised his work and he theirs. Despite discouragement from his father and from Lodge, Ted was considering the idea of giving up doctoring and becoming a professional artist. He still craved proper art lessons. Nevertheless, his 'profession' won out over his 'hobbies', once again and he acquired a post as a locum at the Cheltenham General Hospital. It didn't last long.

In June 1900 Ted received the most astonishing letter from Dr. Sclater, informing him that the post of Junior Surgeon and Vertebrate Zoologist was available for a forthcoming British National Antarctic Expedition and that he would be a suitable applicant. Ted was incredulous.

Discovery with Parhelia, 1905

Discovery
The British National Antarctic Expedition: 1901-1904

"*Ours is a double life, distinct though united.
Our body lives, and our spirit lives; each must be born.*"   EAW

Discovery
The British National Antarctic Expedition: 1901-1904

The British National Antarctic Expedition was part of a co-ordinated pan-European campaign to unveil the frigid enigmas of the Antarctic. Explorers had ventured South before but surprisingly little was known; it wasn't even clear whether a continent existed in the blank space on the map.

Once Ted had been persuaded to apply for the post of Junior Surgeon and Vertebrate Zoologist, he went for it with customary enthusiasm. Uncle Charlie helped where he could and passed Ted's watercolours around the Royal Geographical Society. Ted soon successfully passed his interviews and must have made an impression as he was eventually taken against medical advice and at his own risk. Nowhere else had they managed to find the unique combination of skills which Ted represented. Three weeks before the Expedition sailed, Ted and Ory were married. Their honeymoon was dominated by Antarctic preparations but they were blissfully happy.

Ted was caught up in a whirl of preparations for heading into the unknown aboard the purpose built ship *Discovery*, under Commander Robert Falcon Scott. He brushed up his skills in taxidermy, on the identification of whales and dolphins and read up on subjects such as scurvy. He also designed the Expedition crest for use on crockery and notepaper. Much of Ted's time was spent in the Natural History Museum working on the specimens from the recently returned *Southern Cross* expedition and he wrote the Expedition Report on the seals. Ted's first paper on Antarctic fauna was therefore published before he had ever been South and the illustrations, whilst competent, clearly reflect this. They represent the very style of 'wooden' wildlife art from dead specimens that Ted so wanted to move away from. In this case, however, he had little choice. Upon return, his artwork for the seals in the *Discovery* expedition reports would be intimate portraits of live animals.

BILLY

Discovery sailed from the Solent on 6 August 1901 after an inspection by the King and Queen. Ted watched Ory slowly disappear from view. As they sailed South, he soon found himself to be in his element, working with the other scientists to uncover the mysteries of the ocean. He commenced the large biological collections of the Expedition, collecting and painting seabirds. These became his first serious ornithological watercolours. In addition to painting the specimens, which were destined for British Museums, Ted also enjoyed sitting and sketching birds from the ship in all weathers. He was later to become noted for his artistic achievement in successfully capturing the flight of seabirds on paper, in a period when most artists were still painting seabirds as stuffed specimens. Ted, the amateur dabbler, was by this time inadvertently working at the forefront of modern wildlife painting and seabird ornithology. This isn't to say that there weren't difficulties; he had to work out how to keep his paper dry and to produce an accurate sketch of a seabird whilst the ship was swinging through 30°.

There were challenges in other ways too. Ted found that it took him a considerable effort to adjust to the Navy way of doing things, despite his admiration for several of the officers, including Scott. His most withering criticism, as always, was reserved for those with large egos who didn't pull their weight. He still occasionally let slip a caustic remark and so became known as 'Bill the cynic' to some aboard. However, he viewed his confinement with such men as a part of his ascetic progress: it challenged him to maintain self-control. He gradually earned the respect of his shipmates both for the application of his numerous skills and his diplomacy. As at school and Cambridge, many of his shipmates turned to him as a trusted confidante.

Sailing via Madeira, South Trinidad and South Africa, *Discovery* took a quick excursion to the ice edge off Adélie Land to take magnetic readings and then sailed, via Macquarie Island, to New Zealand. After re-fitting, *Discovery* finally sailed on 24 December and headed for the Ross Sea. Just after midnight on 8 January 1902 they sighted the coast of Victoria Land. From here Ted started on a series of panoramic sketches, producing a unique topographical record of the coastline of Victoria Land from Cape Adare to Ross Island. He also produced numerous sketches of the Great Ice Barrier and the newly discovered King Edward VII Land. During the return exploration along the Great Ice Barrier in search of winter quarters, the first flight in the Antarctic was executed in a balloon. It was made in an attempt to ascertain the extent of the Ice Barrier but no limit could be seen. Ted stayed with his feet firmly on the ground and sketched.

Eventually securing their winter-quarters at Ross Island, *Discovery* settled in for the winter. A series of early sledging explorations were made. The first of these to what would become known as White Island, involved Ted, Lieutenant Ernest Shackleton and the geologist, Hartley Ferrar. It was a sharp learning curve for the 3 men, but they reached the top of the island from where Ted sketched. Many years later the historian, David Yelverton, would show that such a view was not possible from the top of the island, under normal conditions, but only with a mirage throwing distant mountains into view, making Ted's sketch the first known sketch of

HE PREENS!

an Antarctic mirage. Other journeys were also being executed at this time, including an attempt to place a message box at Cape Crozier, at the eastern end of Ross Island. Cape Crozier was a pre-designated message point for relief ships to the expedition. The attempt failed and Able Seaman Vince lost his life in the attempt to return to the ship. Attempts to sledge out onto the Barrier as late as April also ended in ignominy, resulting in a thorough review of their equipment and lessons learned, by Scott and the other officers.

Throughout the winter months, Ted was busy assisting with the wider scientific programmes as well as his own vertebrate zoology work and medical duties. Having become close friends with Shackleton he walked with him to the top of Crater Hill almost daily to read the temperatures at the outlying meteorological station. It was during such walks that he sketched, under extremely difficult conditions, some of the phenomena of an Antarctic winter. Able to work in pencil for only a few minutes at a time, he would then warm his hands in his armpits until the pain of their re-warming had passed, then continue to sketch. Back at the ship he worked up these sketches, along with his many others, into finished pencil drawings and later into rough watercolours and later still into polished watercolours (see for example p36). In this way he secured some extra-ordinary images, including the first accurate images of the *Aurora Australis.* Ted's colour memorisation technique came into its own in the Antarctic. His colleagues marvelled at the accuracy of his work and the sheer quantity of images that he was able to produce. In addition to his more serious artwork, he also substantially illustrated the shipboard magazine, *The South Polar Times.*

Edward Wilson field-sketching during the Discovery expedition

It was during the winter that Scott asked Ted if he would accompany him on the great exploration southward the following season. The journey did not have the Pole as its objective but was an exploration to the South to see what was there, although, of course, no one would have complained if the Pole fell into their laps. Everyone aboard hoped for the chance. Ted was astonished to be asked and not entirely happy to be taken away from his biological work; yet he was quietly pleased to be given such an honour. He persuaded Scott to take one other person and Scott chose Shackleton. Their departure was delayed by a scurvy outbreak aboard which meant that by the time they were ready to depart they knew that there was no chance of the Pole. Ted obtained some notes on scurvy and started to ponder its true cause.

Shortly before their departure on the first Southern Journey, news of an extraordinary discovery arrived back at the ship: the first breeding rookery of Emperor Penguins had been found by a party under Lieutenant Skelton, during their successful visit to erect the message post at Cape Crozier. Skelton had brought live chicks back to the ship. These Ted quickly sketched, creating the first colour images of Emperor Penguin chicks to go with Skelton's recently taken photographs. There was no time for Ted to explore their biology further, as he soon left the ship heading southward.

Instead of finding a flat icy plain all the way to the South, the Southern party came across magnificent new coastline and mountain ranges. These Ted sketched in remarkable detail, often suffering from snow-blindness, or sunburn of the eye, as a result. The optimism of the first part of their journey was soon tempered by the rapid failure of the sledge-dogs and the need to relay their supplies. Ted dissected one casualty to find that it had died from a form of dysentery.

After 60 days of sledging, the men commenced symptoms of scurvy and the dogs were nearly done. It was time to head home. They had achieved a farthest South of 82°. Scott named the inlet at their farthest South for Shackleton and the Cape for Wilson. With the dogs nearly all dead they man-hauled their sledge home. Shackleton was deteriorating rapidly, coughing blood and suffering fainting spells. Scott and Wilson, themselves suffering, struggled to get the party home. It was a close call but they made it. A large dinner was held to celebrate their arrival but the stomachs of the three

Weddell Seals at Hut Point, February 1903. Photograph by Edward Wilson,

hungry men were so shrunken that they spent all night alternately eating and throwing up. Ted noted that he only had a bath after 3 months sledging out of a sense of duty to his shipmates, such was the pain involved in bathing his bruised and sunburned body. Shackleton's deterioration was deemed serious enough to necessitate his evacuation home aboard the relief ship, *Morning*, which had found their winter-quarters thanks to the messages left at Cape Crozier.

As soon as he was able, Ted worked up the many dozens of sketches from their Southern journey and caught up on his zoological work. Topographical pictures, water-colours of parhelia, mock suns, fog bows and other astonishing visual phenomena of Antarctic sledging, started to emerge from Ted's cabin. It was in the execution of these that Ted's study of Turner started to truly come into its own. His application of the technique of halation was used to great effect in recording the boundaries between ice and light.

The Emperor Penguin rookery under the ice cliffs of Cape Crozier, October 1903. Photograph by Edward Wilson

Discovery's second year in the South was an unexpected bonus. The unrelenting ice kept the ship trapped during the summer months and on into the next winter. This gave Scott the unique opportunity to assess the results of the first year's geographic and scientific work and to plan for a second season to fill in the gaps. Ted passed much of the second winter in painting up his pictures, sometimes producing duplicate images of his more popular paintings for competition prizes or friends aboard (see, for example, p46), but he was impatient for the return of the sun. For Ted, this would bring the chance of being the first to study the breeding biology of the Emperor Penguin.

Emperor Penguin, Cape Crozier, October 1903. Photograph by Edward Wilson

The biology of the Emperor Penguin was completely unknown and Skelton's discovery of the first breeding rookery of Emperor Penguins at Cape Crozier represented a major biological coup. Ted had been frustrated not to have the opportunity to study them in depth but the unexpected second season allowed him to do just this. Skelton had found quite well grown chicks in October, so Ted ensured a party left for Cape Crozier early in September, as early in the Spring as he dared, in order to secure good specimens of Emperor Penguin eggs. Instead, to his astonishment, he found quite well grown chicks and was led to conclude that these remarkable birds had to lay their eggs in the middle of the Antarctic winter. It was one of his greatest biological discoveries and one of a series of remarkable field observations that Ted made of these birds over the following months. He led two trips to Cape Crozier during September and October, sketching and watching the Emperors with his naturalist's eye. His field sketches of Emperors fighting over chicks, alive or dead, of feeding, sleeping, locomoting and interacting were considerable and his finished watercolours of the birds much sought after. He also secured numerous specimens, including abandoned eggs, significantly adding to the expedition's biological collections. Towards the end of his final visit he became the first to witness the Emperors migrating out to sea with their chicks on rafts of sea ice.

At the start of December, Ted accompanied a small party led by Lieutenant Armitage across McMurdo Strait and up the Koettlitz Glacier in the area around Mount Discovery. This proved to be his only excursion into the Western Mountains and he relished exploring more of the unknown Antarctic and sketching the mountains up close. It was a relatively short trip as all field trips were ordered to be returned to the ship by Christmas in order to concentrate on freeing *Discovery* from the ice.

Shortly after they had seen in the New Year, Ted went camping with Captain Scott at Cape Royds where he was delighted to find a small Adélie Penguin rookery, which he proceeded to observe and sketch. Scott and Ted were both bewildered to see two ships come into view a few days later. The Admiralty had sent two relief vessels, *Morning* and *Terra Nova* to evacuate *Discovery* in the event of the ice trapping them again. After some days of anxiety, to everyone's delight a swell swept into the Strait and the ice finally freed *Discovery*. No one had wanted to abandon the ship. The Expedition returned up the coast of Victoria Land and then via the Auckland Islands to a triumphant reception in New Zealand. There Ted found Ory waiting for him. They toured the country for several weeks whilst the ship was re-supplied for its voyage home. With Ory safely on her way aboard a steamer, Ted sailed back to England aboard *Discovery*. Sailing via the Magellan Strait, the Falkland Islands and the Azores, she arrived in Portsmouth on 10 September 1904, cheered from every ship and welcomed by huge crowds.

The scientific and geographical discoveries of the British National Antarctic Expedition were considerable. It discovered King Edward VII Land, the Polar Plateau and hundreds of miles of new coastline. It also achieved the first major exploratory sledge journeys to penetrate towards the heart of the Antarctic away from the ice edge. These, along with the considerable collections of geological specimens and fossils all provided convincing evidence of the existence of an Antarctic continent. Further, the magnetic work carried out during the course of the Expedition permitted the location of the South Magnetic Pole to be determined. This was executed as a part of the first major international scientific programme to be carried out in Antarctica and allowed the creation of new maps of magnetism for the Southern Hemisphere. These were of considerable importance to the navigation of shipping in the southern seas. The primary aims of the Expedition were therefore met. In the process further gains were made in numerous other scientific disciplines. Large collections of scientific specimens were obtained, hundreds of new species were described and the understanding of the breeding biology of penguins was commenced. The meteorological work uncovered the existence of the 'coreless' Antarctic winter, one of the great keys to Southern Hemisphere weather patterns. This was so startling that the data was at first publicly criticised by the Meteorological office but it was later forced to retract much of its criticism when it was shown to be in error. The Expedition also determined the nature of the Great Ice Barrier showing that it was afloat, a question that had caused considerable puzzlement to many scientists. In almost every branch of science the expedition returned with new data and from an area considerably further south and under harsher conditions than any of its rivals.

The British National Antarctic Expedition was to be the last in the tradition of the great naval expeditions of exploration dating from Captain Cook. Although it was not officially naval, it was run along naval lines and largely staffed by naval officers and crew. Despite its success, however, the expedition had in many ways simply whetted appetites for further exploration and research, making it the crucible of the Heroic Age of Antarctic Exploration. All such further expeditions would be private enterprises.

Ted was proud to have been a part of this great national achievement, although the heroic role, which the adoring British public tried to thrust upon him, was not one which he much relished. Not least amongst the results of the expedition was an extraordinary visual record of the Antarctic, due to his tireless work aboard. His pictures attracted a good deal of positive comment when they were exhibited.

Opening at the Bruton Street Gallery in London the exhibition of the expedition's achievements then toured the country. In particular, Ted's paintings of Emperor Penguin chicks started a long love affair with these birds in the heart of the British public. Even such noted figures as Sir Joseph Hooker, who had been South with Captain James Ross aboard *Erebus* and *Terror* (1839-1843), commented upon the quality of his pictures and especially of his seabirds. Ted was rapidly becoming a household name.

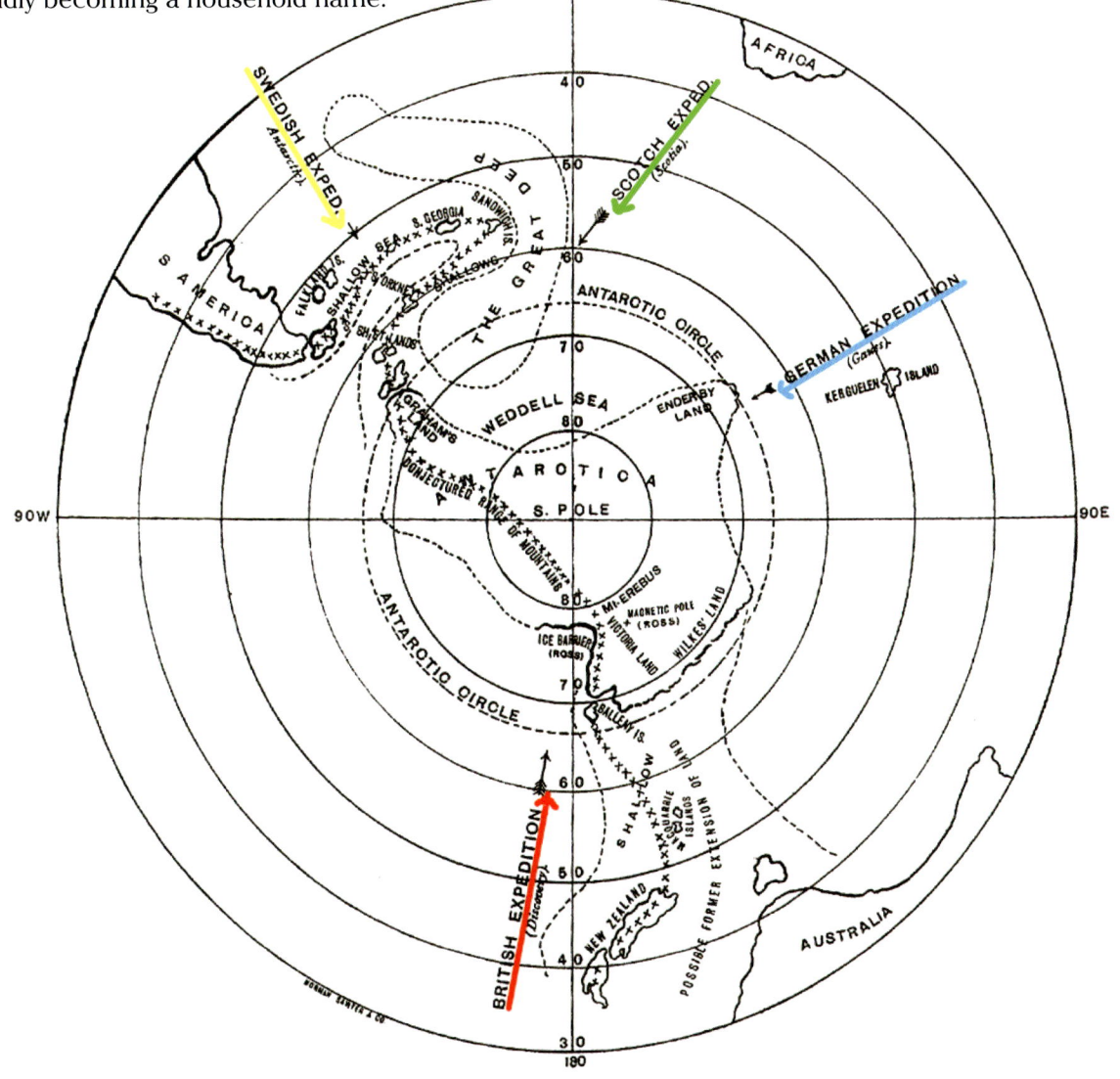

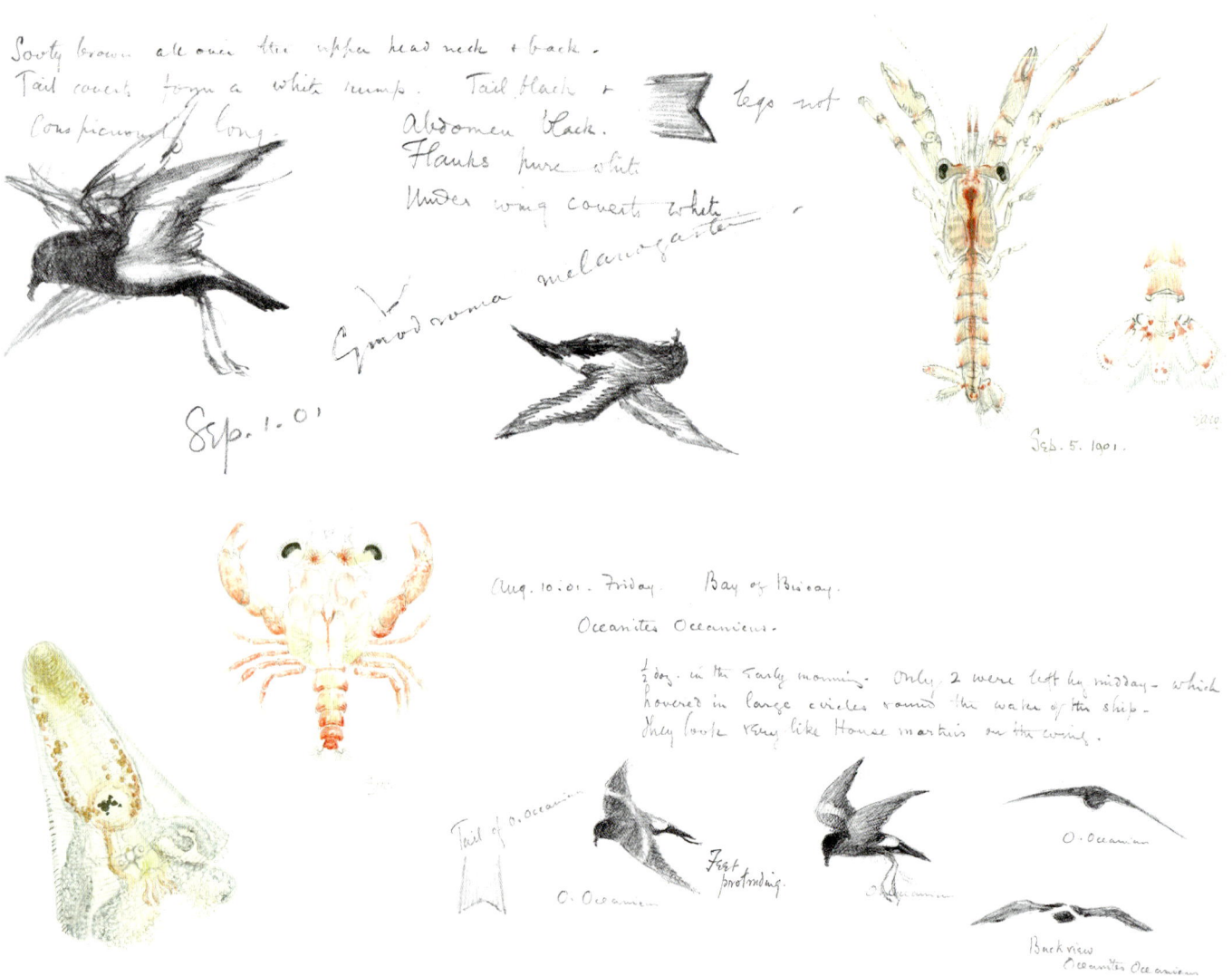

The British National Antarctic Expedition: 1901-1904

Heading South: August - September 1901

South Trinidad Petrel - sitting - Sep. 13. '01. EAW

"The sunset was even more glorious than the sunrise, for the sky was almost cloudless and we got the intense yellow ochreous glare after sunset uninterrupted by any clouds. It was almost uncanny. One felt as though something terrible was about to happen - the same sort of feeling that one gets in dense yellow London fog, only this was beautiful, and magnificent, as well as terrifying. Everyone was on the bridge watching it."
EAW

South Trinidad - Monument - Sugar Loaf - & Noah's Ark - Sep. 13. '01. E.A.W.

The British National Antarctic Expedition: 1901-1904

Heading South: September - October 1901

South Africa, October 1901

"And they call it a "dry" ship - that's what beats me. Why every 6 or 7 minutes as I write, an egg cup full of cold water comes down on my page or my head or my neck as the case may be, and that is inside a tight deck-light and goes on in every cabin in the ship."
EAW

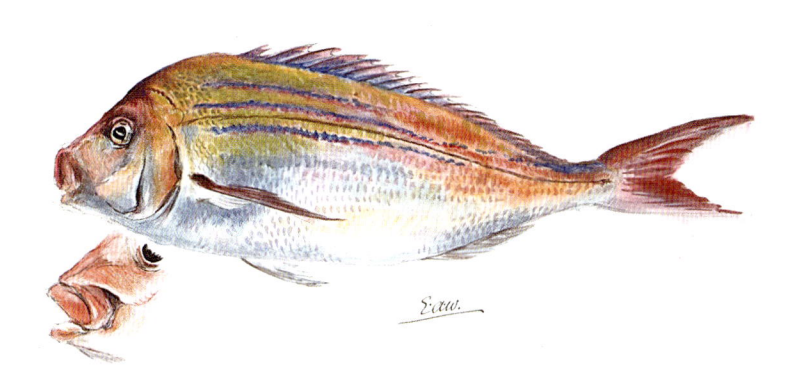

The British National Antarctic Expedition: 1901-1904

Heading South: October - November 1901

Nov. 17. 1901. Pack ice pieces. 3 to 5 feet high. Ice bases + compacted snow above. North of Adélie Land

Prion banksi. Skin No. 7.

EAW.

The lamellae are seen only at hinder part of the gape. Even when the bill is closed they are visible there but not elsewhere.

Thalassoeca antarctica. Pack ice. Nov. 1901. EAW.

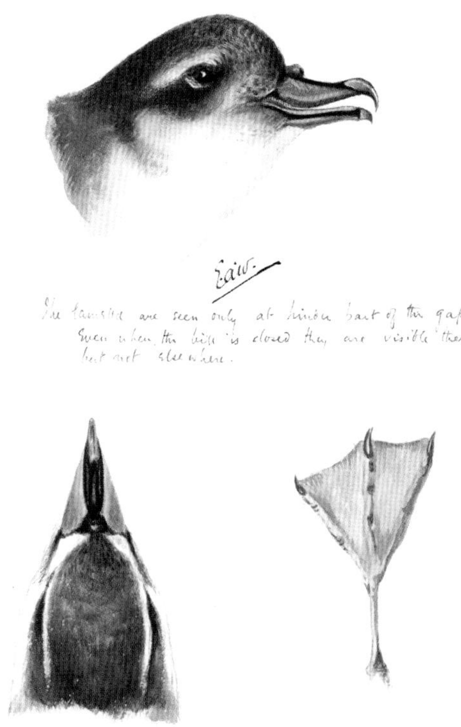

To show colour of head + foot of Skin No. 9 during life. Nov. 16. 1901. in the pack ice 61°46'S. 141°12'E. "Discovery" Antarctic Expedition.

"And now we had loose ice all round us and here and there great frozen hummocks, where slabs the size of kitchen tables were thrown one on the other anyhow and so frozen, with every hollow and crack and crevice a perfect miracle of blue and green light, and then came the ice birds - the Southern Fulmar, a beautiful bluish grey bird the size of a Common Gull, and then the black and white Antarctic Petrel, dressed like a domino, and then to crown all the Snow Petrel, pure white like a slim fantail, with black eyes and beak and the most graceful flight. With us into the ice came the Sooty Albatross and Prion petrels and Halobaena, and Cape Pigeons." EAW

Nov. 17. 1901. Small tables of ice + compacted snow. Not more than 3 to 4 ft high

The British National Antarctic Expedition: 1901-1904

Macquarie Island, November 1901

- 26 -

Heading South: November - December 1901

New Zealand and Southern Ocean

"And then a few seconds after, one of these enormous waves broke right over the ship. We all three hung on to the stanchions and rails and were swept clean off our feet. We were simply deluged, and I burst out laughing at the Skipper who was gasping for breath. He had been nearly a minute under water, and then I saw the whole upper deck was afloat… Didn't we just laugh, and weren't we jolly wet. Our cloths were soaked through, all down our necks as well and our sea boots were filled." EAW

The British National Antarctic Expedition: 1901-1904

Greyheaded Albatros.
Thalassogeron culminatus.
Dec. 29. 1901. ♀
(Skin. N° 6.)

E.A.W.

What a mouth, what a mouth
what a mouth by South

Diomedea melanophrys.
Dec. 28. 1901.
E.A.W.
(Skin. N° 1.)

Phoebetria palpebrata

Phoebetria palpebrata

Heading South: December 1901 - January 1902

Jan. 1. 1902. A school of these porpoises played round our bows for a long time today.

"Truth is like a lighted lamp in that it cannot be hidden away in the darkness because it carries its own light." EAW

View from above.

The second Berg sighted by the "Discovery" on her way South from New Zealand. Jan. 4. 1902. E.A.W.

Entering the Pack Ice

(Skin. No. 15.)

(Skin No. 28)

The British National Antarctic Expedition: 1901-1904

Heading South: January 1902

Changes in the feet of Pygoscelis adeliae, the Adélie Penguin.

"About 5pm we at last rounded Cape Adare and could see past two rocks (the 'Sisters') the flat triangle of shingle on which stood the huts of the Southern Cross expedition and some millions of the Adélie Penguin. Such a sight! There were literally millions of them…The place was the colour of anchovy paste from the excreta of the young penguins. It simply stunk like hell, and the noise was deafening."
EAW

Cape Adare

Cape Adare Penguins.

Young.

The Two Sisters - Cape Adare.

The British National Antarctic Expedition: 1901-1904

Heading South: January 1902

Cape Adare

"We found several Skuas on the way up, with their fledglings, pretty greyish white fluffy things with pale blue bill and feet. But the old birds kept up a continual attack on us when we were anywhere near their young, chattering excitedly and dashing at our heads so boldly and persistently that we had to protect ourselves." EAW

The British National Antarctic Expedition: 1901-1904

Cape Wadworth, Coulman Island

Midnight Sun on Admiralty Range. from off Cape Adare. Jan. 10. 1902. E.A.W.

Cape Anne, Coulman Island

Heading South: January 1902

"All this in glorious sunshine and a brilliancy of light and colour, almost a monochrome in these months of midnight sunshine. For one sees little but white and blue, the blue sometimes drifting into green, sometimes into purple or lilac, but always very tender and beautiful. All this makes up a scene one cannot but wonder at, and rejoice in, it is so very glorious." EAW

The British National Antarctic Expedition: 1901-1904

Great Ice Barrier: January 1902

"The colours at midnight seem to me to be getting more beautiful. The ice floes are constantly bright lilac or pink, the water in between reflects the yellow or the bright green of the sky and the shadows in the ice hummocks are as usual pale sea green and blue, the purest colours on earth." EAW

The British National Antarctic Expedition: 1901-1904

- 38 -

Great Ice Barrier: January - Febuary 1902

Jan. 28.02.

Eastern end of Great Barrier,

Pieces of Barrier ice.

Jan. 28. 1902. Stratified pieces. white compacted snow
Off King Edward's Land. 6 to 10 ft above water.
78° 24' S. 167° 56' W.

Jan. 28. 1902. Stratified pieces. white compacted snow.
6 to 10 ft above water.
78° 24' S. 167° 56' W.

Feb. 7. 1902. Lat. 77° 08' S. Long. 168° 30' E.

Berg recently tilted - showing incrustation of ice formed at previous sea level.

"There was a good deal of fog and mist, giving one the most beautiful effects of light and shade and mystery in the ice all around us. The sea was full of heavy ice… these bergs were in every stage of demolition and tilted up at every imaginable angle… I spent the day skinning and drawing." EAW

The British National Antarctic Expedition: 1901-1904

"Discovery" in Winterquarters looking West.

Hut Point. "Discovery's" Winterquarters. 1902-1904.

Arrival at Hut Point - Winterquarters: February 1902

View from Crater I. on Island I. looking S and SW.

"*Then we had time to look to our frostbites. Shackle's ear was badly blistered and his hand was still all white. We rubbed it till it had got painful and then it blistered and was very sore the next few days.*" EAW

First explorations: White Island, February. 19-22 1902

The British National Antarctic Expedition: 1901-1904

? *Berardius arnouxi.* Beaked Whales. Eight or ten, sporting & breaching in Terror Bay.
  Duvernoy.   Feb. 25. 1902 - "Discovery".

Sunset looking across McMurdo Sound to the Royal Society Range

Winterquarters: February - March 1902

"The cold has a funny effect on pencil lead, the softest B is a hard and gritty as an H and makes the same sort of mark. You can't get any soft black mark with it out of doors." EAW

View from Hut Point Peninsula to Winterquarters and the Royal Society Range

The British National Antarctic Expedition: 1901-1904

Across McMurdo Sound. Views of the Western Mountains from Discovery in Winterquarters: the Royal Society Range with Mt. Lister.

Winterquarters: April 1902

"Imagine our delight too on seeing here also a group of 15 splendid Emperor Penguins in the pink of condition. We made our way to them and they came to meet us with evident signs of interest, and though they objected to being stroked and immediately tobogganed off on their stomachs if we tried to stroke them, they always came back at once and stood up again to enjoy our company. They were really most wonderfully handsome - bright lemon yellow all down the front, rich black heads, blue backs and rose pink and lilac beaks." EAW

The British National Antarctic Expedition: 1901-1904

- 46 -

Antarctic Winter: May - June 1902

"Along this rope we make our way to the screen which is on the floe. Now on a night such as this, the rope gets buried in snow drift in less than 2 hours, though it was four feet off the ground, and the snow is so compact that you can't drag the rope up. Neither can see one post from the other. So one of us goes ahead to find the next post and having found it he stands there till you come to him, and so on in turns, till you both reach the screen smothered in snow from tumbling all over the place, half blind and with a thin layer of ice all over your face from melting snow drift. All this time you have been nursing a slow match wick, which is all you have to read the thermometers and the wind scales with." EAW

"ON THE WAY TO THE SCREEN IN A BLIZZARD."

The British National Antarctic Expedition: 1901-1904

The Aurora Australis - Antarctic Winter: 1902

"Taking the observations as usual at 8am I saw a fine aurora, a long low arch of vertical streamers from S. to E. and N. It was dead calm, beautiful still and clear and silent, and these long straight beams of light all round the horizon lend a weird charm to the darkness and the silence." EAW

"The sketch is intended to give an idea of the culmination of moving arcs and curtains and beams of light in the zenith, where occasionally the appearance of radiation from a centre, or of revolution, slow and changeful, about an axis, is given to the onlooker. Such evanescent figures, almost impossible to describe in words, are even more difficult to suggest with any approach to truth, in pictures; they are spoken of as coronae." EAW

The British National Antarctic Expedition: 1901-1904

The Gap in winter moonlight

Antarctic Winter: May - June 1902

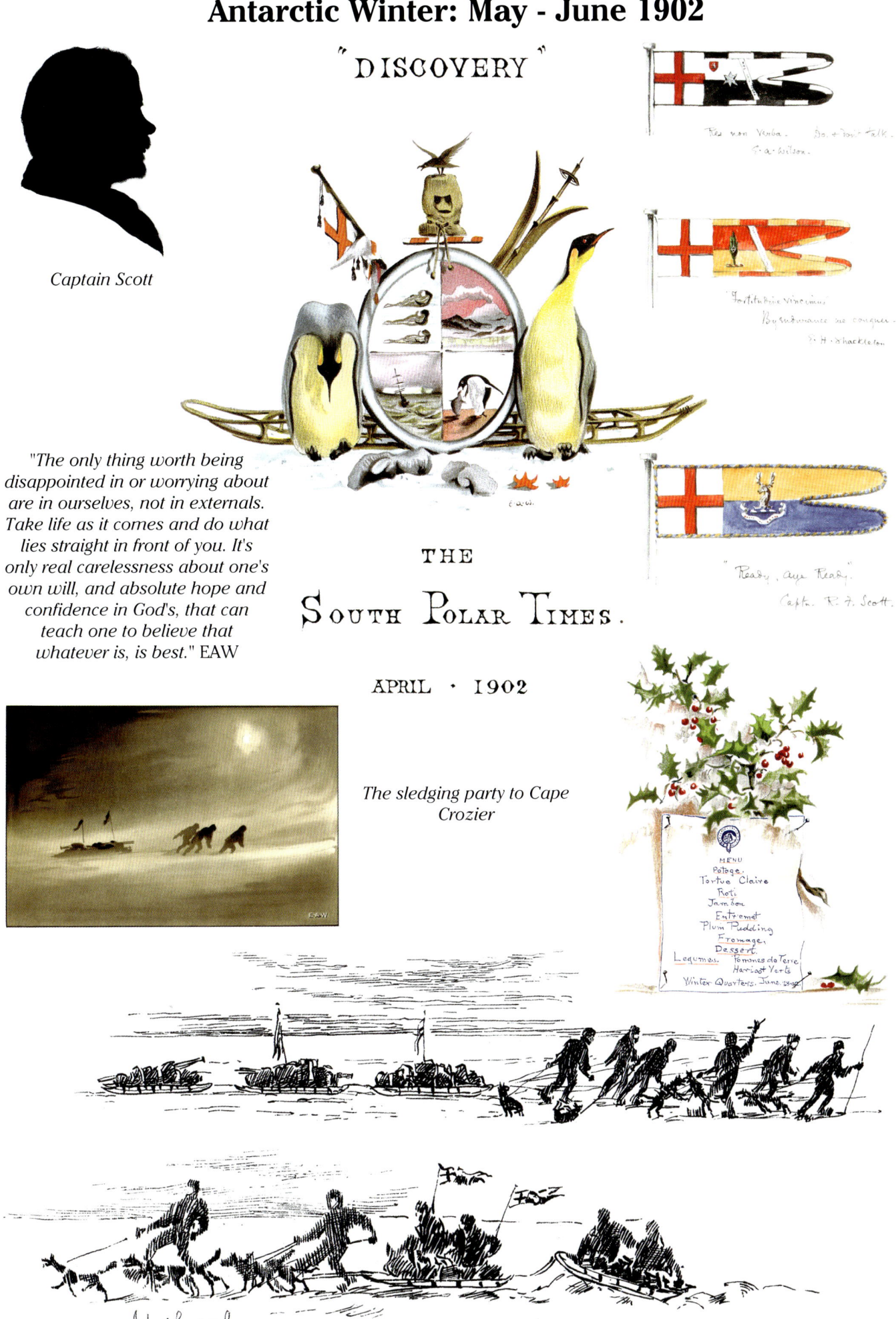

"DISCOVERY"

Captain Scott

THE SOUTH POLAR TIMES.

APRIL · 1902

"The only thing worth being disappointed in or worrying about are in ourselves, not in externals. Take life as it comes and do what lies straight in front of you. It's only real carelessness about one's own will, and absolute hope and confidence in God's, that can teach one to believe that whatever is, is best." EAW

The sledging party to Cape Crozier

- 51 -

The British National Antarctic Expedition: 1901-1904

Fireglow on the smoke of Mount Erebus

The Return of the Sun: July - September 1902

"The mixture of lights and the delicacy of all the colouring is most extraordinary. I think perhaps, after all these months of darkness, one's eyes appreciate the faintest differences in tints more than ever. I have never seen anything like them before. I only wish I could reproduce them. I tried till dinner time when I came in." EAW

Royal Society Range with Mt. Lister - Western Mountains, McMurdo Strait. August 14 1902, midday

Looking north in McMurdo Strait: July 26 1902, midday

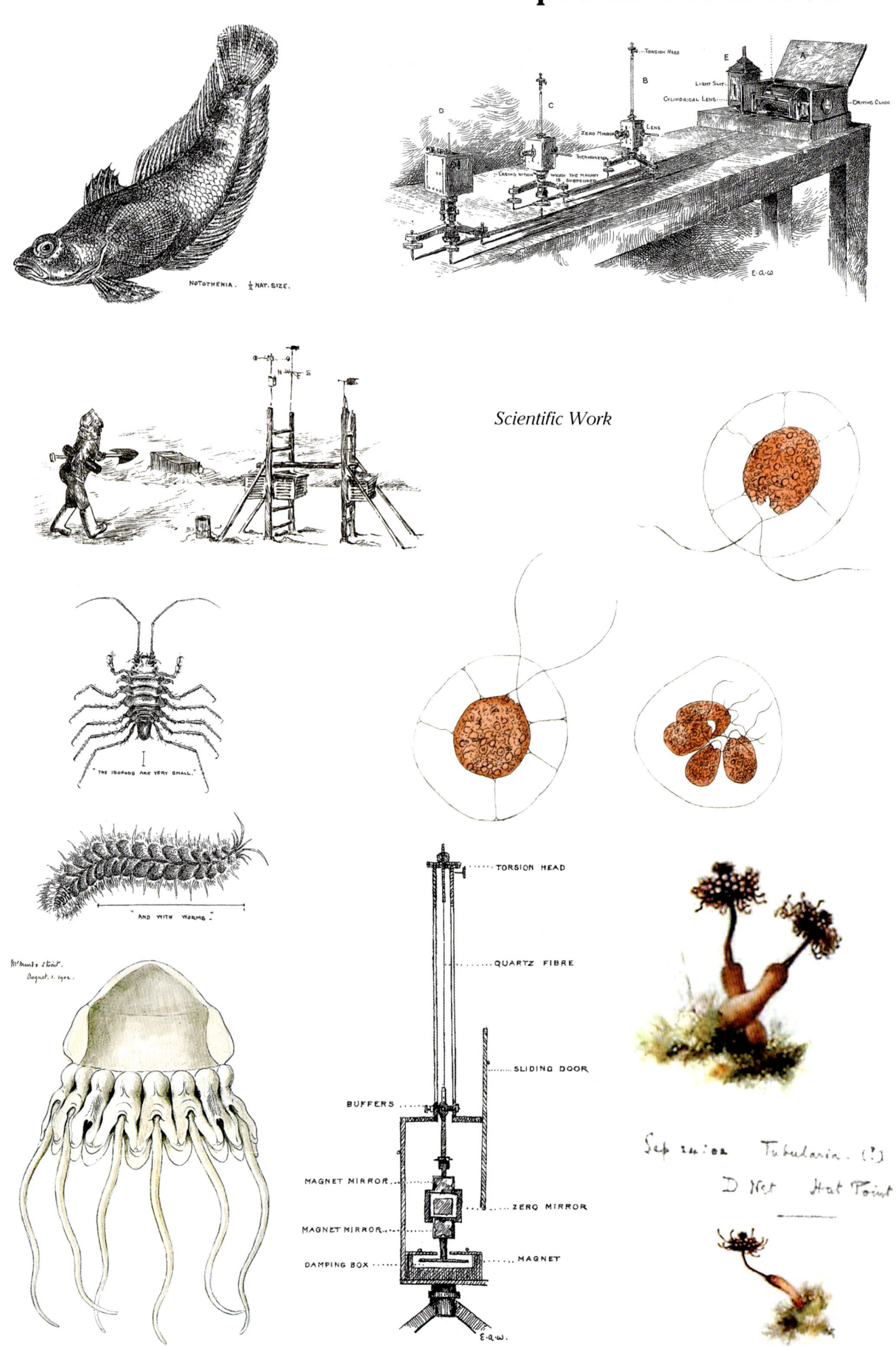

The British National Antarctic Expedition: 1901-1904

Scientific Work

The Return of the Sun: July - September 1902

"The lights this evening baffle all description. We had Erebus all ablaze in orange sunlight just above us, Terror aflame with a red light all down one side a long way further off, the sky all pink and deep cobalt with a wonderful afterglow and everything else in pure blue and violet shadow… The sunset itself was a blaze of gold and the whole sky from north to south was strung with long streaks of rosepink clouds. It was a very wonderful sight with the contrast of pale emerald ice cliffs, white and lilac snow and black volcanic rock." EAW

Setting a Fish Trap.
Mt Erebus & Mt Terror from the Turtle Back Island.

THE BIOLOGIST AT WORK IN HIS SHELTER.

The British National Antarctic Expedition: 1901-1904

Looking towards Crater Hill, early August 1902, midday

McMurdo Sound, sun's approach, August 1902

The Return of the Sun: August - October 1902

"They are very handsome too and of a colour I had not suspected. The back is covered with a very soft white fluffy down which deepened into grey on the underparts. The forehead, head and neck were jet black and a large patch including eye, ear and cheek was the purest white. A most striking contrast. The beak was white and black and the eyes, as far as I could make out, were of the bluish opalescent colour that characterizes the eyes of many fledglings." EAW

The first paintings of Emperor Penguin chicks, October 1902

The British National Antarctic Expedition: 1901-1904

Antarctic Sledging: 1901-1903

"Our furs are simply awful. It takes ever so long and quite exhausts one getting into them. They are frozen stiff as boards and once in, one can do literally nothing but lie as one falls in the tent. One cannot turn, and one has a longing to turn over on the other side the whole night through, a thing which is utterly impossible without upsetting both your companions and making yourself and them uncomfortable for an hour or so. Your only chance of keeping warm is to lie still."
EAW

The British National Antarctic Expedition: 1901-1904

The Southern Journey: October 1902 - February 1903

"*We saw a very wonderful exhibition of mock suns, parhelia, perihelia and circles of light in the sky connected with the sun in a blue sky, with showers of ice crystals flying about, just for all the world like spring showers of rain at home as one sees them from the Crippetts. This was such a striking sight that we at once halted and got out the theodolite and took all the angles and elevations and bearings of the various circles and mock suns. It was a wonderful and very beautiful sight but very hard to describe, but I got sketches which I hope will give some idea of it. I had never seen anything like it before. It beat all the halos and parhelia we have ever seen at the ship.*" EAW

Joe.

Bismark

Jim.

- 61 -

The British National Antarctic Expedition: 1901-1904

Farthest South, December 28-31 1902 Mt. Longstaff & Cape Goldie

The Southern Journey: October 1902 - February 1903

"My eyes have been in a sorry state all day from sketching with sunglare, streaming with water and very painful from time to time. Sketching in the Antarctic is not all joy, for apart from the fact that your fingers are all thumbs, and are soon so cold that you don't know what or where they are, till they warm up again in the tent (<u>then</u> you know all about it!); apart from this you get colder and colder all over, and you have to sketch when your eyes stop running, one eye at a time, through a narrow slit in snow goggles. No one knows till they have tried it how jolly comfortable it all is." EAW

Farthest South, December 28-31 1902 Mt. Markham, Shackleton Inlet & Cape Wilson

The British National Antarctic Expedition: 1901-1904

The Last of the Dogs

Fog-bow – on the Great Ice Barrier

Winterquarters: February - March 1903

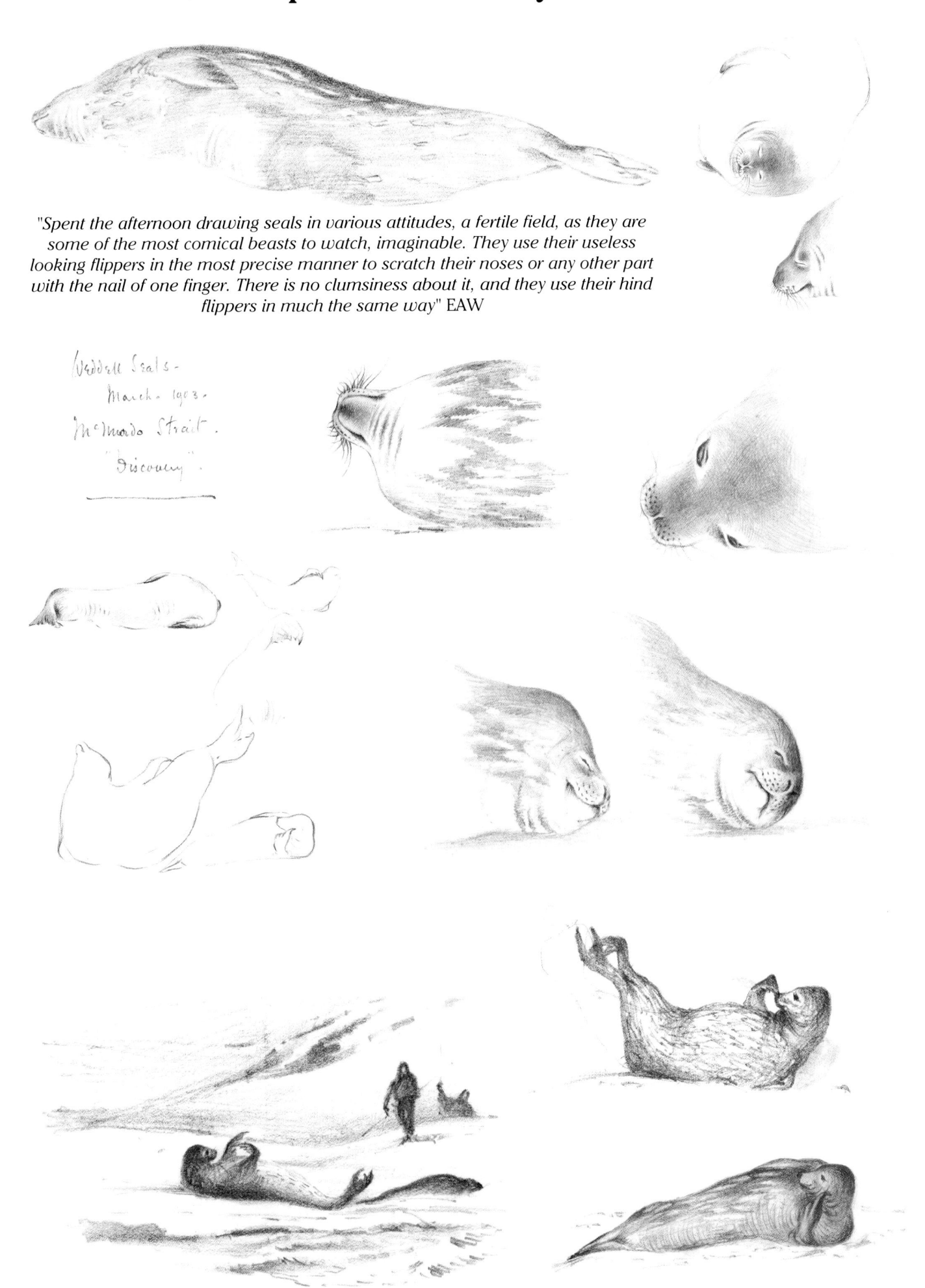

"Spent the afternoon drawing seals in various attitudes, a fertile field, as they are some of the most comical beasts to watch, imaginable. They use their useless looking flippers in the most precise manner to scratch their noses or any other part with the nail of one finger. There is no clumsiness about it, and they use their hind flippers in much the same way" EAW

Weddell Seals -
March - 1903.
McMurdo Strait.
"Discovery".

The British National Antarctic Expedition: 1901-1904

Discovery in Winterquarters, looking North

Winterquarters: March - April 1903

"I went up the hill taking my ski and had a good run along the top of the promontory to Castle Rock which I then sketched. Though the temperature was between 20 and 30 below zero, it was so calm that I sketched for a good half hour with bare hands."
EAW

Castle Rock from the Ski Slopes, looking NW

Crater Hill from Harbour Heights looking SE

The British National Antarctic Expedition: 1901-1904

Earth Shadows

Winterquarters: March - April 1903

Mt Erebus, from Crater Hill,
showing the usual form taken by the smoke - bulky masses
tailing off into a long pennant of some twenty miles very often.

April. 26. 1903.

"The sky was simply beautiful, and Mount Erebus with its roll of smoke was lit up with beautiful pink and lilac opal tints against a sky of pale yellow. The sun's direction was marked by a long bright glow of rose pink, fading upwards into orange and green and blue. The stars were brilliant and the southern sky a deep ultramarine blue."
EAW

The British National Antarctic Expedition: 1901-1904

The Aurora Australis

Antarctic Winter: May - June 1903

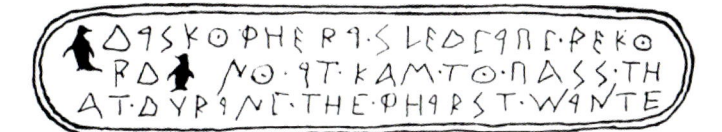

WE WISH ALL OUR READERS
A
MERRY
MIDWINTER.

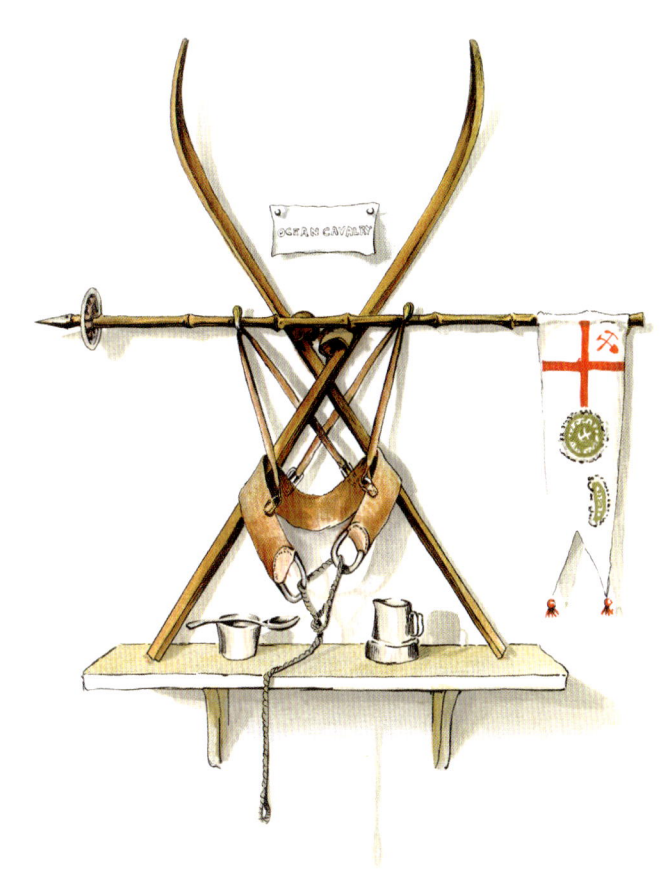

"How intense is the delight of puzzling out a thing for oneself… No one can know who hasn't tried. It fairly makes one hug one's joy as it dawns on one" EAW

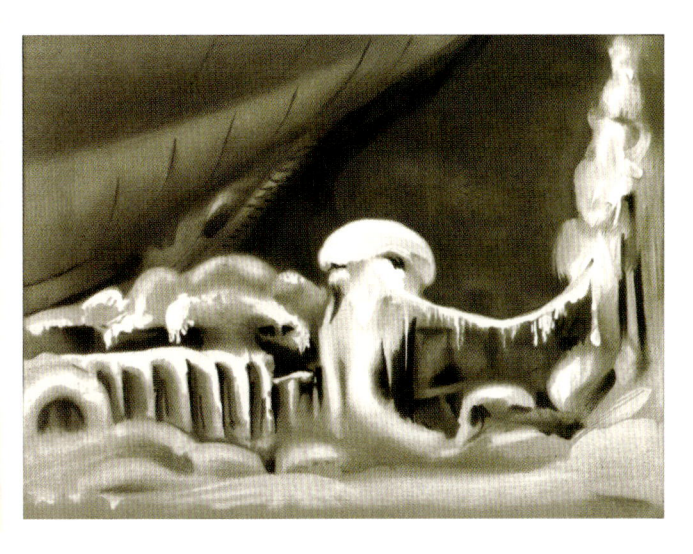

The British National Antarctic Expedition: 1901-1904

Discovery in Winterquarters looking north

Looking NE about noon, Mount Terror, Castle Rock and Danger Slopes, July 20 1903

The Return of the Sun: July - September 1903

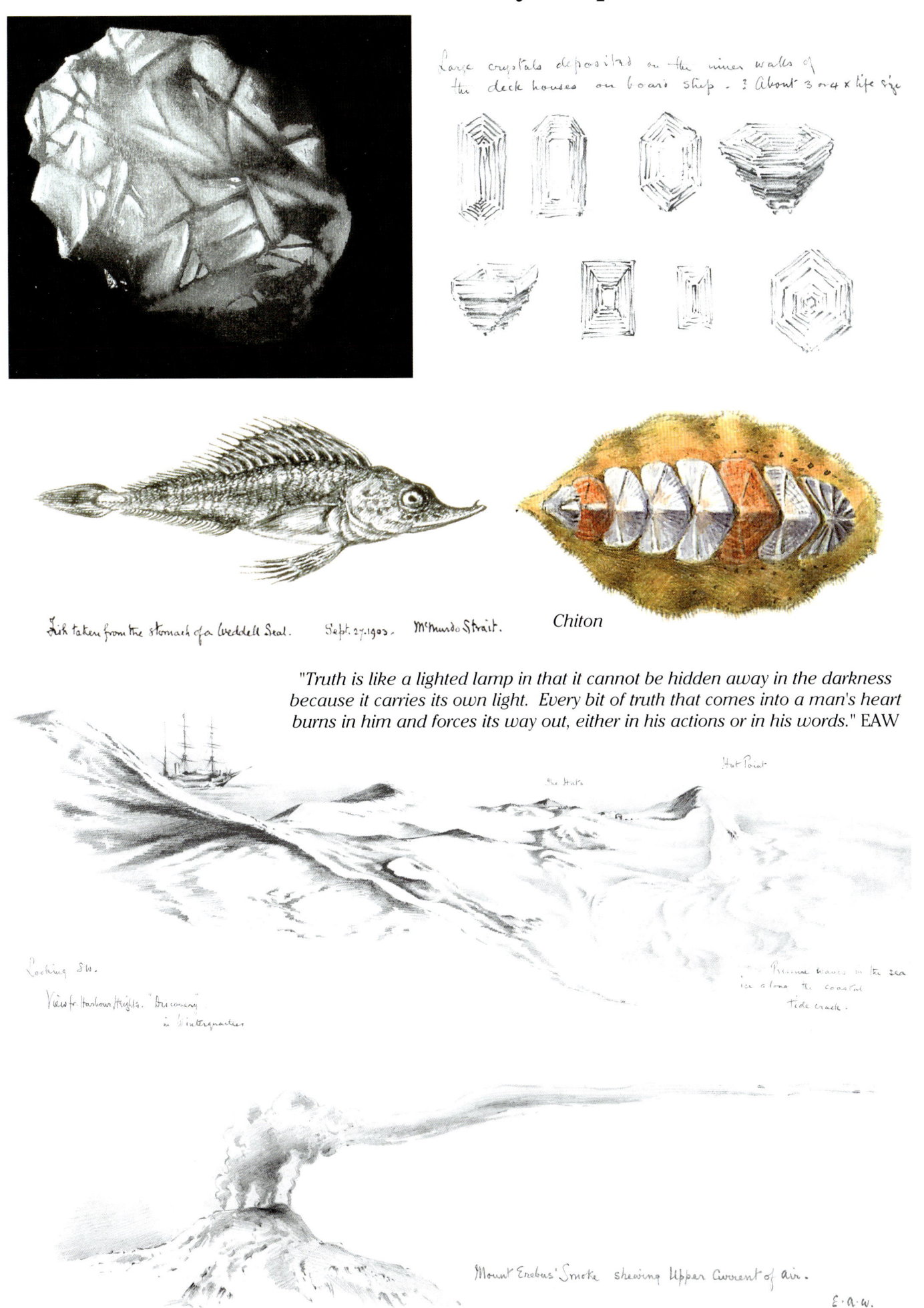

Large crystals deposited on the inner walls of the deck houses on board ship. ? About 3 or 4 × life size

Fish taken from the stomach of a Weddell Seal. Sept. 27. 1903. McMurdo Strait.

Chiton

"Truth is like a lighted lamp in that it cannot be hidden away in the darkness because it carries its own light. Every bit of truth that comes into a man's heart burns in him and forces its way out, either in his actions or in his words." EAW

Looking SW.
View fr. Harbour Heights. "Discovery" in Winterquarters.

Hut Point
The Huts
Pressure waves in the sea ice along the coastal Tide crack.

Mount Erebus' Smoke shewing Upper Current of air.
E.A.W.

The British National Antarctic Expedition: 1901-1904

South Polar (McCormick's) Skuas near nest

Discovery in Winterquarters, September 13 1903

The Return of the Sun: July - September 1903

*Opalescent Alto Stratus and snow drift,
looking North, August 17 1903
1-2pm*

"*In the north at noon there was a splendid sunrise with a heavy bank of cloud arranged for all the world like wavy hair, and wherever the sunlight caught these waves and curls, it was broken into the most delicate opal or mother of pearl tints, all colours of the rainbow - pale rose, pure lilac, emerald green, lemon yellow and fiery red, all blending one with another, but with no apparent arrangement. So that a wisp of cloud, standing like a stray curl in the blue sky, would be lit by pink and brilliant lilac and then would begin to shine at one end with a light that can only be compared with the light you see in a vacuum tube with a current sparking through it, or perhaps the colour is more exactly what you get with incandescent barium.*" EAW

*Looking West to Mt. Lister,
McMurdo Strait,
August 16 1903 2pm*

*Looking South, Mt. Discovery,
August 3 1903 1pm*

The British National Antarctic Expedition: 1901-1904

Cape Crozier: September - November 1903

Cape Crozier
Sep. 13. 03.

Emperors fighting over a live chick

"I submitted myself to much vicious pecking and many slaps from the parents' wings in feeling under them for the chicks to see in what position they were carried, and I found the position a happy go lucky one. For the chicks lay in any attitude, even on its side across the parents' feet" EAW

Emperor Penguin Chick. sleeping
Cape Crozier - Sept. 1903

Emperor approaching dead chick
intending to pouch it.

Emperor Penguin,
attempting to drag a dead chick into its lap, in its desire to brood over something.
Cape Crozier rookery. Sep. 13. 1903.

Emperor Penguin,
showing the method it has of holding the young chick.
Cape Crozier rookery. Sep. 13. 1903.

Emperor Penguin,
feeding a chick.
Cape Crozier rookery. Sep. 13. 1903.

The British National Antarctic Expedition: 1901-1904

Cape Crozier: September - November 1903

No. 13. *Aptenodytes forsteri* -
Emperor Penguin - chick. just hatched. life size.
Drawn in position in which it was found frozen at the
rookery at Cape Crozier - Sep. 13. 1903.

No. 5. Emperor Penguin, chick - life size.
Preserved in formalin. *Aptenodytes forsteri*.

Drawn in the position in which it was found frozen at the
Cape Crozier rookery - Sep. 13. 1903.
Shows the bare patch of skin on the abdomen, above the anus.

"It was a very great surprise to us to find that all the old ones had hatched their eggs and were already nursing chicks apparently every bit as old as those were that Royds and Skelton brought home the year before, on October the 18th. Over five weeks later than we are now. I cannot understand this at all, for though among so many birds it is impossible to be certain that none of them were still incubating eggs, we certainly saw nothing but well grown chicks under all that we examined." EAW

The British National Antarctic Expedition: 1901-1904

Cape Crozier: September - November 1903

Antarctic Summer: November - December 1903

"Three or four miles up this valley was a second Chinese wall, an abruptly ending glacier which ended in a cliff, 300 ft. above sea level. All the coast here was a collection of moraine deposit, with many pools of water among the moraine heaps, full of a red and green shiny algae, the only vegetation to be found, which formed a brown paper sort of high-water mark round the edges on the stones. I got a sketch of this glacier tongue while Armitage and Heald searched out a path among the moraine heaps…"
EAW

Koettlitz Glacier, Western Mountains:

The British National Antarctic Expedition: 1901-1904

Cape Royds, January - February 1904

Heading North: February - March 1904

Memorial Cross to Seaman Vince

The relief ships, Terra Nova and Morning, heading home

"I shall not forget the last view we had of our straight as we last saw Mount Discovery in the midnight sunset. It was a blaze of colour astern of us, a glowing orange fire and standing out black on our starboard quarter was the Terra Nova under sail and farther astern on our port quarter, the Morning, with all our sympathies, for all that as a relief ship she was not much use" EAW

The Possession Islands

The British National Antarctic Expedition: 1901-1904

Mt. Minto and Mt. Adam, February.24 1904

Coast of Victoria Land

Southern Cross Hut, Cape Adare, February 25 1904

Heading North: February - March 1904

Shot in Laurie Harbour. ♂
Port Ross. Auckland Isles.
March 23. 1904.

Nesierax novaezelandiae
(Skin N° 107.)

(Skin No. 120.)

Sterna frontalis.
Laurie Harbour. Port Ross.
Auckland Isles. March 25. 1904.

Platycercus auriceps. Kakariki.
Yellow fronted Parakeet.
Laurie Harbour. Port Ross.
Auckland Isles. March 26. 1904.

Quail Falcon. ♀ (Nesierax novaezelandiae.)
Shot in Laurie Harbour. Port Ross.
Auckland Islands.
March 16. 1904.
(Skin N° 108.)

E.A.W.

Auckland Islands

"A short life with hard work
is so much preferable to a
long one with comfort"
EAW

Shot in Laurie Harbour. & Enderby Island.
Port Ross. Auckland Islands.
March 20. 1904.

The British National Antarctic Expedition: 1901-1904

Strait of Magellan

*Heading North, March - September 1904:
New Zealand to England*

- 86 -

Illustrations for the Scientific Reports of the British National Antarctic Expedition: 1901-1904

Antarctic (Discovery) Exp.   Birds, Plate III.   E A Wilson pinx / Bale & Danielsson Ltd lith

Various stages in the plumage of McCormick's Skua (Megalestris McCormicki).
1 & 2. both nestlings.
3. Fledgling, not quite able to fly.
4. Young adult. 5. Old adult.

"Principles are the laws of life which each person makes for himself, and the best people are those whose principles are so strong that they resist every temptation to anything lower, yet so pliant that they readily give way to anything higher. Like a cog-wheel with a catch, they can always be screwed a turn higher and never drop to where they were before." EAW

Antarctic (Discovery) Exp.   Seals, Plate I.   E A Wilson pinx / Bale & Danielsson, Ltd lith

Red Grouse

The Natural History of the British Isles: 1904 - 1910

"The keynote of all my longing is to know as much as I can about anything that is still as God made it; I never have the slightest feeling that country rambling is mere amusement or waste of time, it is converse with God through His works. That sounds almost too fine to be true." EAW

The Natural History of the British Isles: 1904-1910

The return of the British National Antarctic Expedition to the United Kingdom was greeted with wide acclaim and an almost endless series of dinners. For the first time in his life Ted found fame, which he disliked and shrank from at every opportunity. Nor did he enjoy the commercialisation of his work. He had innocently agreed to produce copies of his paintings for a modest price out of deference to the Royal Geographical Society which had put up the Expedition. Over 80 orders were placed which Ted diligently executed at his home in Bushey over the next 6 months, but he received a good deal of trouble and little of the 20-30 guinea picture price (approx. £1500-£2000 today) for his efforts. Whilst his illustrative process led from rough field sketch to finished pencil sketch to rough watercolour and then finished watercolour, he often threw the rougher versions away, once completed, leading to few multiple images. He often threw away finished watercolours, also, repainting them if he thought they were not good enough.

Mr and Mrs Reginald Smith, with Oriana Wilson and Captain Scott, picnicking near Cortachy, c.1907. Photograph by Edward Wilson

Similar pictures were sometimes generated when he painted the same views in changing lights, creating a series of studies. However, it was a new experience for him to need to paint multiple 'fair' copies of the same picture. He had started it aboard ship as gifts or prizes to shipmates but now found himself executing multiple 'originals' of the same image for a demanding public. It meant that, uniquely to this expedition, there were multiple variations of his more popular paintings. Many of these pictures have circulated on the art market ever since. Ted disliked every minute of it.

Ted also undertook to write and illustrate the reports on birds and mammals for the *Discovery* Expedition Reports. This resulted in being employed for four days a week at the Natural History Museum, working with eminent zoologists on the Expedition collections. Ted wanted his monograph on the Emperor Penguin to become a classic. He also undertook work for books connected with the Expedition, in particular Captain Scott's own book The Voyage of the *Discovery*, a facsimile of The South Polar Times and the Expedition's Album of Photographs and Sketches. It was 1908 before he had completed the numerous publication and illustrative projects that arose from the Expedition. These projects helped to keep him in regular contact with Captain Scott and introduced them both to Mr. and Mrs. Reginald Smith of the publishers, Smith, Elder and Co. They were to become close personal friends. His work was to bring him in contact with many of the leading scientists and artists of his day. Sir Joseph Hooker had made his admiration for Ted's work widely known. They later met and Ted studied Hooker's Antarctic portfolio. He was also introduced to the artists Herkomer and Swan, with Ted subsequently becoming friends with many of the artists at the Herkomer School near to Bushey. Co-incidentally, this is where he and Ory had set up their new home. He executed a particularly striking series of charcoal portraits of horses at the Herkomer School, which by this point was run by Lucy Kemp-Welch, a noted equestrian artist.

Ted and Ory occasionally managed to escape his increasing work load to spend time with his family in Cheltenham. Although his mother had by now given up the lease on *The Crippetts*, Ted still liked to walk up to the farm and to visit its woods and hedgerows. It was a special pleasure that Ted liked to indulge in and which became rarer as his success grew. Apart from anything else, Ted was much in demand as a lecturer and delighted his audiences with tales and impressions of penguins. He also began to campaign for the protection of penguins, which were being boiled for their oil in increasing numbers. He gave papers on the subject before the Royal Society for the Protection of Birds (RSPB) and the International Ornithological Congress. It started a process which finally met with success many years after Ted's death.

It was after addressing the British Ornithologists Union at the Restaurant Friscati in March 1905 that Ted was introduced to Lord Lovat. The leading ornithologist, Ogilvie Grant, had invited Lord Lovat to attend the meeting especially to meet Ted. Lord Lovat was the Chairman of the Board of Agriculture's Commission on the Investigation of Grouse Disease. The Board of Agriculture had commenced the Inquiry because grouse shooting was an important part of the rural economy and the birds were dying of a mysterious disease. The cause had so far eluded researchers. Lord Lovat

Edward Wilson c. 1910

was looking for a new field naturalist who was also a bacteriologist and doctor to carry out the field-work for the Inquiry. This contract was expected to engage the person concerned for six months of the year. Ogilvie-Grant thought that Ted was the perfect candidate. Upon meeting him, Lord Lovat was so impressed that he invited Ted to Carlisle the following day to a meeting of gamekeepers who were to discuss the subject. Ted agreed and after attending the Carlisle meeting accepted the post of Field Observer to the Commission.

Additional offers of work kept pouring in. Of particular excitement was contact from an old Cambridge friend, Gerald Barrett-Hamilton, who was a keen admirer of Ted's work. Barrett-Hamilton had been engaged to produce a new edition of Bell's *A History of British Mammals* and he wanted Ted to do the illustrations. Ted jumped at the chance, the thought of illustrating many of the animals of the British countryside, which he knew so well, was hugely appealing. Shortly after he had agreed to this Ted was contacted by the ornithologist Eagle Clarke who had been commissioned to produce a new edition of Yarrell's *A History of British Birds*. He, too, wanted Ted to provide the illustrations. Ted was overjoyed; the chance to illustrate all 350 species of British birds was a project that was so close to his heart that he simply could not refuse. Since the grouse work was only scheduled to take up six months of his year, Ted was confident that he could meet all of these other demands in the six months remaining to him. He was determined that every species of mammal and every species of bird in Britain would be illustrated from live field sketches, rather than from stuffed specimens, making these projects ground breaking initiatives.

Picnic plate - 'Before'

Before these projects began in earnest, however, a plan had been hatched by Ted's parents for a full family holiday in the West of Ireland. The Wilson family had a tradition of such holidays, during which the time would be spent largely in picnicking, reading, sketching and natural history collecting. It was to be the last such family holiday, although this wasn't realised at the time. Despite his parents instruction that it was to be a total holiday, such were the demands on Ted that he nevertheless took unfinished work with him to complete.

Ted's work as the Field Observer to the Grouse Disease Commission soon started in earnest. By early 1906 there were dead grouse "pouring in by every post" to the Wilson home at Bushey. Every dead grouse found on a British moor was dispatched to Ted for examination and a good many were suffering from delays in transit. Over the course of the Inquiry he would dissect just under 2,000 birds, leading one of his colleagues to comment that in whatever location Ted was situated he was constantly surrounded "by a halo of grouse feathers and unravelled entrails". This was not terribly popular in station hotels. As he travelled around the country in pursuit of his work, odious remains were occasionally forgotten. The vast majority of the grouse sent to him he turned into museum skins, as well as taking detailed biological notes. Every bird then had a report filed upon it. Upon filing one such report on a brace of grouse recently received through the post, Ted received the astonished reply that they had been intended for him to eat, not to dissect. In all of this work Ted received the competent support of Ory, who was, to all intents and purposes, his unpaid assistant.

In July 1906 Ted and Ory moved up to Scotland to be nearer to the grouse work. Lord Lovat put his shooting lodge at Glendoe at their disposal and they frequently stayed here during the Inquiry. The Reginald Smiths also made their small cottage at Cortachy available to the Wilsons and it was Cortachy that became their home away from home.

The Grouse Disease Inquiry was taking ever increasing amounts of Ted's time and far more than the six months per year that he had been led to expect. He often had to travel from one end of the country to the other for meetings or to visit grouse moors, the owners of which often entertained him so well that it "cuts into my work". He often spent his train journeys catching up on proof corrections or correspondence. He had no time for reading and so, often as not, Ory would read out loud to him whilst he was dissecting grouse. With the expansion of the grouse work Ted was seriously over-committed and as a result became severely over-worked. Yet he dealt with the many demands upon him with a typically calm assiduousness, doing each and every thing in front of him without fuss and to the best of his ability. He was determined to explain the cause of the grouse disease and equally determined that each and every plate for the mammals and the birds should be a unique composition based, wherever possible, on his personal observation.

Picnic plate - 'After'

Despite the fact that grouse were continuing to arrive in the post, none had yet arrived with the dreaded disease. In fact it took 3 years before an outbreak of the disease occurred again. Ted's suspicion as to the cause of grouse disease had started to fall upon a minute threadworm which he had found infecting the cœca of weaker birds and appeared to interrupt the digestive process, so killing the bird. It was of considerable delight to Ted, therefore, to discover the worm alive and wriggling in a diseased grouse lung, to watch them hatching and burrowing in the gut and to find dead worms in the liver. He was getting close to being able to prove that the threadworm was indeed the cause of the grouse disease. Ted rapidly traced the mode of infection to the dewdrops on the tips of the young shoots upon which the grouse feed. It took him many early mornings of study upon the moors to gather the evidence for this but he revelled in it. Ted was rarely indoors except when examining grouse or working with his co-workers upon the microscopic examination of organisms. They were searching for the secondary host, in order to give a complete account of the life of the threadworm and the course of the epidemic disease. However, the secondary host proved somewhat elusive. Controlled experiments at Cambridge allowed Ted to prove the causes of the disease, however, and the modern management of Grouse Moors is still carried out in accordance with the recommendations arising from his work.

"DISCOVERY" ANTARCTIC EXHIBITION

(By kind permission of Sir Clements Markham, K.C.B., and Captain Scott, R.N.)

WILL OPEN AT THE

Bruton Galleries

13, BRUTON STREET, BOND STREET, W.

ON FRIDAY,

NOVEMBER 4th, 1904.

AND WILL INCLUDE

WATER COLOR SKETCHES and COLORED DRAWINGS by

Dr. Edward A. Wilson,

AND

PHOTOGRAPHS taken by

Lieut.-Engr. Skelton, R.N.

A Model of "THE DISCOVERY"

KAYAKS,

SNOWSHOES,

SLEDGE FLAGS,

AND OTHER ARTICLES OF INTEREST USED IN THE SOUTH POLAR REGIONS.

The EXHIBITION will be on view daily, 10 to 6
(including Saturdays).

The Bruton Galleries, 13, Bruton Street, Bond Street, W.

In February 1907 an unexpected letter arrived from Shackleton informing Ted that he was heading back to the Antarctic with his own expedition and with the objective of reaching the South Pole. He wanted Ted to be his second in command. Ted declined, due to the fact that he was already over-committed and really did not feel that he could leave the grouse work. Ted also wrote to Lord Lovat re-affirming his commitment to seeing the work through. Shackleton failed in his attempts to extricate Ted from the Inquiry. These letters were followed several weeks later by letters from Scott, who was upset that Shackleton had announced that he was going to use his old base without consulting him. Scott was also brewing plans for another expedition. Both men turned to Ted to mediate in their dispute and a deal was brokered. Shackleton returned from the Antarctic in June 1909, a hero for getting within 98 miles (176km) of the Pole, but he had failed to attain it. He had, however, broken his agreement with Scott and for all that he was now hailed as a hero, Ted nevertheless broke off their friendship. He thought that Shackleton had 'dragged the name of polar exploration into the mud of his own selfish ambition'.

Moonlight on a Frozen Sea. Plate for Scott's The Voyage of the Discovery, 1905

Scott rapidly got his dormant plans under way and Ted was publicly confirmed as the Chief of Scientific Staff. Such was the public outbreak of 'Pole mania' those plans now had to publicly include the Pole but Ted hoped to make bagging the Pole 'merely an item in the results' of a substantial geographic and scientific programme. Ted hoped that this position would help him to get a regular billet on his return, so that he could settle down with Ory, perhaps in his long-hoped for role as a Government scientist in New Zealand. Many of Ted's friends tried to dissuade him from going South again but he was already committed. His friendship and loyalty to Scott were strong and they were determined to finish together what they had started aboard *Discovery*. Many of the plans for the expedition were fermented at the Reginald Smiths' cottage in Cortachy whilst Ted was studying the grouse.

Ted was now in serious difficulties with regard to meeting all of his work commitments. He was busy painting and writing at all hours of the day. It was a relief, therefore, when the publishers agreed to suspend the project for the British Birds provided that Ted met the contract for the first half of the plates before the Expedition sailed. This Ted willingly agreed to do and so it was with some astonishment that he greeted the news that the contract was terminated with immediate effect some weeks later. The precise reasons for the cancellation are unclear, all sorts of reasons were given at the time, but the project never saw the light of day, neither with Ted nor anyone else as illustrator.

The project for the British Mammals, however, was a considerable success. Published in parts from 1910-1921, *A History of British Mammals* by Barrett Hamilton and illustrated by Edward Wilson became a classic work on the British mammals. It is still sought after and was republished as recently as 1978. This is perhaps especially surprising since it was never completed and many of Ted's plates were never used. In particular, Volume III on the marine mammals, was never produced. Whilst Ted had completed all of the required plates, Ted wanted to re-paint those illustrating the cetacea, as he considered them to be inadequate. However, both the author and illustrator died within a few months of each other and it seems most likely that this was the reason for the eventual incompletion of the project. The first part was issued in October 1910, shortly after Ted had sailed South aboard *Terra Nova* so he never saw the finished work. He was sending the last illustrations back for the work some months after sailing.

Ted was also trying hard to finish writing up the grouse work before heading South. Of the final Report of the Commission, Ted wrote approximately one third and provided the majority of the illustrations. A paper promised to the Zoological Society on the changes of plumage in Red Grouse, was written on the train and finished in a station waiting-room. He was so tired that he took to working standing up, so that he didn't fall asleep over his work. Ted was still trying to finish the grouse work when he sailed South in June 1910. The final instalments were posted home to the Commission from South Africa in August. The grouse work was completed at last. Ted was also destined never to see the results of his work in the final Report of the Commission. However, the Grouse Disease Inquiry stood for almost a century as the most extensive scientific investigation carried out into a disease in a wild bird population. It was the founding of modern field ornithology. For his part in it Ted was later recognised as being amongst the world's top ornithologists for the first part of the Twentieth Century.

Ted had little time to 'go the rounds' and say farewell to his family. The Wilson family, including Ted and Ory, gathered briefly at the family home at *Westal* in Cheltenham for what was to prove to be the last such family Christmas. On top of his other work, Ted was drawing up a scientific programme for the forthcoming Expedition, along with finding the scientists to fulfil it and the equipment that they would need. The greatest challenge of his career lay before him.

Valencia Island from Anascaul, Ireland, August 1905

Lunar Corona, Cape Evans, 1911

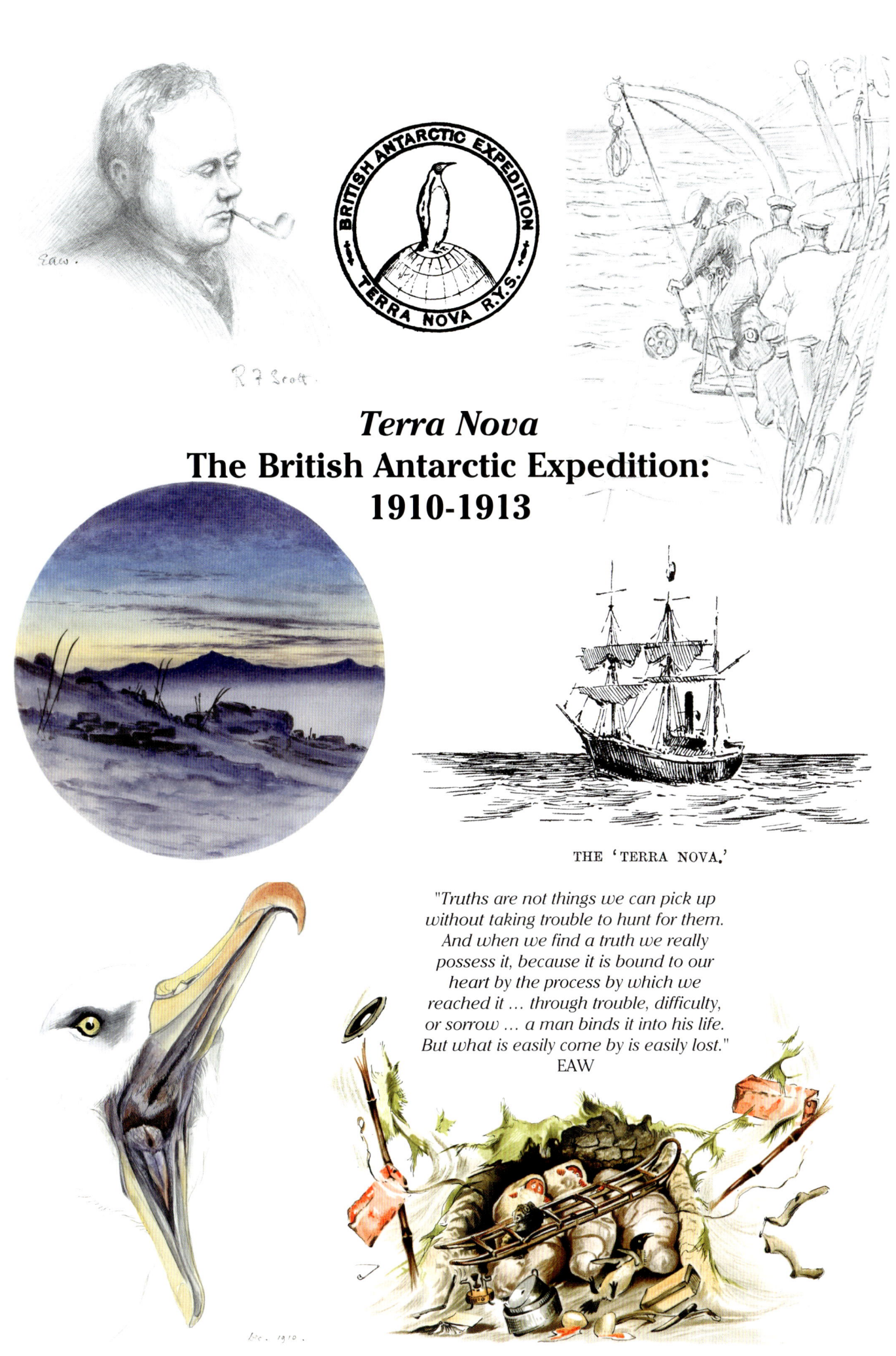

Terra Nova
The British Antarctic Expedition: 1910-1913

THE 'TERRA NOVA.'

"Truths are not things we can pick up without taking trouble to hunt for them. And when we find a truth we really possess it, because it is bound to our heart by the process by which we reached it ... through trouble, difficulty, or sorrow ... a man binds it into his life. But what is easily come by is easily lost."
EAW

Terra Nova
The British Antarctic Expedition: 1910-1913

Amid a cacophony of hooters and cheering crowds, *Terra Nova* sailed from Cardiff on 15 June 1910, bound for the frozen South. Sailing via Madeira, South Trinidad, South Africa and Australia to New Zealand, Ted was sketching all the way. At South Trinidad, he was able to see, once again, the Trinidad Petrel that had been named for him by the ornithological authorities, but his keen field observation cast doubt upon its status as a species. From South Africa he posted home the last of his mammal and grouse work. At Cape Town too, Captain Scott joined the ship. He had stayed behind in London to raise badly needed funds for the Expedition. Scott now sent Ted on ahead to Australia, aboard the RMS *Corinthic*, to carry out fund-raising there. Fund-raising was an almost constant headache. The entire Expedition finally assembled in New Zealand from where it sailed on 29 November 1910. Despite having received a telegram from the Norwegian explorer, Roald Amundsen, informing Scott that he was "sailing South", there was still no real comprehension aboard as to what this truly meant. In fact, Amundsen had deceived everyone as to his true intentions and was deliberately setting out to forestall the British at the South Pole.

After fierce storms and a severe delay in the pack ice, the Expedition finally arrived at Cape Crozier, where it was intended to locate the expedition base. Cape Crozier provided excellent access to both Adélie and Emperor Penguin rookeries and also an easy route onto the Great Ice Barrier, which formed the road to the Pole. However, on the day of their arrival a heavy swell precluded any form of landing and with the expedition already behind schedule and running short of coal due to the delays, it was decided to head into McMurdo Sound and locate Winterquarters there instead. A suitable spot was soon located and re-named Cape Evans, for Lieutenant Evans, Scott's deputy leader. Within hours they were effecting a landing and had soon established their base.

Edward Wilson birdwatching aboard Terra Nova, 1910 by Raymond Priestley

By 24 January the first major sledging parties were underway including a party heading South to establish supply depots towards the following season's attempt on the Pole itself. This Depot Journey proved to be a catalogue of bad luck and human error which left the Expedition's main supply depot, One Ton Depot, 30 miles further north than had been intended and with the loss of six ponies. Several of these were lost in an incident on the sea ice when trying to reach their old *Discovery* hut at Hut Point. The dogs had not performed well either. Scott had intended to use his dogs to get to the South Pole but during the return journey an entire team fell into a crevasse. Scott insisted on being lowered personally in order to effect their rescue and was never heard to talk of taking dogs to the Pole again. He no longer believed that he would get the dogs up the Beardmore Glacier with its huge crevasses successfully. These crevasses had proved the ruin of Shackleton's attempt on the Pole when he had lost his last pony and a sledge of supplies into a dark abyss. Into the middle of these concerns arrived the news that Amundsen had been found, having set up his winter quarters on the Great Ice Barrier at the Bay of Whales. This was a tremendous shock; in Antarctic terms he was 'just next door'. Additionally, his presence had seriously disrupted Scott's scientific and exploration programme, rendering Scott's second exploration team, the Eastern Party, implausible, thereby turning it into a Northern Party. The Northern Party would instead explore the area around Cape Adare and later around Mount Melbourne. After numerous arguments and much fretting, Scott decided to simply stick to his plans as they were laid out, rather than engaging in the race that Amundsen had declared. He was not prepared to abandon his scientific effort in order to pick up Amundsen's gauntlet.

Wilson and Pennell salting seal skins aboard Terra Nova, Dec. 27 1910, by Herbert Ponting

For several weeks all returning sledge parties were lodged at Hut Point whilst waiting for the sea ice to form and allow a safe return to Cape Evans. Scott was frustrated, but there was nothing much to be done about it. It was during these weeks at Hut Point that Ted was able to relax a little and he completed a large number of sketches. They finally returned to Cape Evans on 21 April.

Through all of these events, Ted was a bastion of calm support to Scott. Indeed, he became the confidante of almost all on the Expedition, who regarded him with considerable affection, calling him "Uncle Bill". They trusted his common-sense judgement, his personal serenity and his complete dedication to others. What they didn't know was the considerable personal effort that Ted had made, over the years, to achieve such self-mastery, nor that the basis of this achievement, lay in a strong Christian faith. He quietly organised and inspired the scientific staff to carry out an extraordinary amount of work as well as executing a great deal himself. He was forever working at his table studying biological specimens or painting up his sketches, frequently standing so as not to fall asleep. In particular, during the first months of the winter Ted executed some of his finest watercolours. It is this series of his paintings that are perhaps most widely known. The accuracy of his Antarctic work in terms of colour and detail has frequently been noted, in particular by Apsley Cherry Garrard. He later wrote of these paintings that "If you look at a picture of a parhelion by Wilson not only can you be sure that the mock suns, circles and shafts appeared in the sky as they are shown on paper, but you can also rest assured that the number of degrees between, say, the sun and the outer ring of light were in fact such as he has represented them." Nevertheless, Ted was not that pleased with them. The acetylene lamps, which provided the light in the hut throughout the winter darkness, cast a yellow tinge to his colouration and he determined that every picture would need to be re-painted in order to ensure complete perfection of colour. It was this level of scientific accuracy, also developed with painstaking self-discipline over many years, that made Ted an important expedition artist. Also with the team at Cape Evans was the first professional photographer to head South. Herbert Ponting was aboard at Scott's personal invitation to ensure that the photography on this expedition would make a serious contribution to the scientific work. Ponting would not disappoint. His filming of the precise technique that a Weddell Seal uses to make its breathing hole in the ice gave him particular pleasure as it disproved all the existing theories, including Ted's. All photography in the Polar Regions is a footnote to Ponting's achievement. Just as Hodges had once crystallised the form of exploration art whilst working with Cook, so Ponting crystallised the form of polar photography whilst working with Scott, and he defined the visualisation of the continent for the next century. Film and photograph, rather than art, were, from this point on, the primary media for recording expeditions. With the last of the great expedition artists and one of the first great travel photographers working side by side, this expedition is unique in its vast pictorial record. Ted and Ponting got on famously and assisted each other with criticism and inspiration. They planned to exhibit their work together upon returning home and it seems astonishing that this was never achieved.

Edward Wilson after sledging, by Herbert Ponting

Apart from his painting and regular scientific work, Ted had something rather unusual planned for the winter programme. The fact that they had been unable to make their base at Cape Crozier meant that a winter journey needed to be undertaken in order to secure the eggs of the Emperor Penguin at an early stage of embryonic development. One of the great scientific questions of the day was the search for evidence to test the hypothesis of Darwin's Theory of Evolution, published 50 years earlier. One proposed solution was that, with the absence of proof in the fossil record, it might be found through embryology. The hypothesis

Bowers, Wilson and Cherry-Garrard, departing for the Winter Journey to Cape Crozier, June 1911, by Herbert Ponting

'Our Bill', cartoon of Edward Wilson for the South Polar Times by Lillie, Sept.1911

proposed was that ontogony (the development of the individual) recapitulated phylogeny (the evolutionary development of the species). Since it was widely believed that the Emperor Penguin must be a primitive bird, it was thought that the study of the embryo of this species would probably yield the scientific Holy Grail; the missing evolutionary link between dinosaurs and birds. The prize could not have been higher, even though it now meant making a journey of considerable difficulty. Scott tried to dissuade Ted from going, but it is probable that Ted had made completing this work a pre-condition for his agreeing to return South and Scott, for all his fears, would not break his word.

Additionally, however, a winter journey gave an unexpected opportunity to answer questions that were of critical importance to the success of the forthcoming Pole journey. Firstly, it provided the chance to try out different sledging rations under extreme field conditions, with varying quantities of fats etc. to ensure that they had the very best sledging rations for the Pole journey ahead; and secondly, Scott had requested that his meteorologists deliver predictions for the most likely weather conditions to be encountered on the Great Ice Barrier during his attempt on the Pole. In order to complete these to their satisfaction, they needed weather data from the Great Ice Barrier in winter. Thus the Winter Journey now became of utmost importance to Scott and the success of the Pole Journey, as well as being of the greatest scientific interest.

Ted and his companions, Birdie Bowers and Apsley Cherry Garrard, left Cape Evans on 27 June 1911, marching out into the winter darkness for Cape Crozier, 65 miles (120km) away. The average temperature on the journey was -60°F (-51°C); it fell as low as -77°F (-60°C). It was so cold that the pus inside their blistered frostbite froze and their teeth cracked, but they persevered. At Cape Crozier they built a stone hut to serve as a field laboratory to study the penguins. However, shortly after they had attained the first 3 eggs they became trapped by a blizzard. Their tent blew away, shortly to be followed by the hut roof. As they lay in their sleeping bags with the blizzard raging around them, Ted passed his 39th birthday. Incredibly, when the blizzard was over, they searched and found their tent. Birdie Bowers tied himself to it whenever they stopped during the return journey, so that it didn't get away without him again. The three men arrived safely back at Cape Evans on 1 August.

Ted himself called the Winter Journey "the weirdest bird's-nesting expedition that has ever been or ever will be". He was disappointed not to have seen more of the Emperors and to have collected so few eggs, but no one doubted the extraordinary courage the men had shown in this greatest of scientific quests.

The comparative study of sledging rations was also considered to be a success. In six weeks, under the harshest of conditions, the men had perfected the balance of rations and lost little body weight. The rations for the Pole journey were organised accordingly. However, no one knew in those days that a higher calorie intake was needed at altitude (calories and vitamins were unknown) and so the field-tests, even in extreme conditions, are subsequently known to have been inadequate. Nevertheless, it explains Scott's confidence that they had done everything scientifically possible to ensure the best sledging ration yet devised.

Likewise the meteorological data obtained during this journey allowed for the completion of the meteorological projections upon which Scott finalised his plans for his assault on the Pole. In the event they were not to be the weather conditions that were encountered and explains Scott's surprise in his diary at the weather conditions. For years the meteorological projections were questioned as a result, along with Scott's comments on the weather. It took the work of Dr. Susan Solomon in her book *The Coldest March* to finally show that Scott's meteorologists had got their projections right - and that Scott genuinely did encounter unusual weather patterns.

The main Pole Party left Cape Evans on 1 November 1911. Following his experience on the *Discovery* expedition, Scott's assault on the Pole utilised several modes of transport: motor sledges, dogs, ponies and man-hauling. The route to the pole from Cape Evans was in four stages: first was the short preamble across the sea ice to the *Discovery* headquarters at Hut

Edward Wilson with his pony, Nobby, Oct. 1911, by Herbert Ponting

Point; the second across the Great Ice Barrier to the Trans-Antarctic Mountains; the third was the haul from sea level up to 9000 feet (2743 m), ascending the great Beardmore Glacier; and the final stage across the high-elevation Polar Plateau. The return journey would be along the same route, picking up depots of supplies laid by the supporting parties during their outward marches, for a total journey of 1566 miles (2900 km).

The outward leg of the first two stages, from Cape Evans to the Beardmore Glacier, was where Scott chose to deploy his transport. He doubted that any team of men could man-haul the whole distance to the Pole and back, but the animals and motorised sledges could give them a good start. The final two stages, up the Beardmore Glacier and then on to the Pole would be made by man-hauling, as would the entire return journey to Cape Evans, since Scott did not believe that any other mode of transport could negotiate the

The Norwegian Tent at the South Pole, Wilson sketching, Jan.1912, by Birdie Bowers

conditions on the Beardmore. Unlike the dogs, the ponies would be shot for food and it was hoped that fresh meat for the men would help to avoid scurvy, still then an unexplained disease. The dogs were to be saved for future scientific work. Scott also hoped that his motor sledges would prove themselves over the Great Ice Barrier and whilst they did not perform as well as had been hoped their main difficulty was in the air-cooling system. Nobody thought they would need anything more sophisticated in the Antarctic and the engines overheated. Scott was nevertheless delighted with them and was writing to his partners to ensure that the patents were in order during the traverse of the Barrier. Scott was right. His motor sledges were the future of polar transport and also proved to be the forerunners of the tanks of the First World War.

From the beginning, the Southern Parties encountered unusual weather, particularly in the form of a warm wet blizzard at the base of the Beardmore Glacier. It slowed them down and sapped their strength. Despite this, they more or less maintained their schedule, frequently stopping to sketch, photograph and collect scientific specimens. This was not a simple dash to the Pole but the execution of a pre-planned scientific programme. Ted's notebooks are full of instructions as to what specimens, sketches and features required the particular attention of himself or other scientists on the march.

At the top of the Beardmore Glacier Scott announced the final team for the Pole, with 5 men rather than 4, amongst which Ted was delighted to be included. The team of Scott, Ted, Bowers, Oates and P.O. Evans achieved the South Pole on 17 January 1912. Here they found Norwegian flags and a tent; Amundsen had beaten them to it. It is doubtful if Ted was very bothered. He dutifully stood and sketched; they took their positions, posed for photographs and headed north again a couple of days later. It was on the way back down the Beardmore Glacier that things started to go badly wrong. Evans was becoming noticeably weak. They stopped to collect geological samples in the area around Mount Buckley, as they had been requested to do by senior geologists, and picked up 35lbs (16kgs) of specimens. Amongst these was the first known Antarctic specimen of *Glossopteris*, a critical specimen in the debate on the origin of the Earth's continents. Ted made detailed geological notes and sketches as they descended the Beardmore. They were to be his last. His time was increasingly taken up nursing sick companions. Evans died at the bottom of the Beardmore Glacier; Ted thought that he had injured his brain in a series of falls. The remaining four men soldiered on, but once on the Ice Barrier they experienced unseasonably low temperatures. Instead of the expected temperatures around -20°F (-29°C) they experienced temperatures down to -40°F (-40°C). These altered the surface of the snow and made it almost impossible to pull their sledge. One by one they became frostbitten, Captain Oates so badly that he committed suicide in a bid to save his companions, walking out of the tent with the words "I am just going outside and may be some time". It was not enough. Badly frostbitten, dehydrated and short of food and fuel, Ted, Bowers and Scott perished in their tent around 29 March, 11 miles (20km) short of their main One Ton Depot. Scott noted that Ted was cheerful to the end. The following Spring they were buried upon the Great Ice Barrier by a search party, which also recovered their scientific specimens, their diaries and the final sketch books of Edward Adrian Wilson.

Wilson, Scott, Evans, Oates and Bowers at the South Pole, Jan. 18 1912 - the last known photograph taken by Edward Wilson

The British Antarctic Expedition: 1910-1913

N.E. Trades →

July. 6. 10. "Terra Nova".
? Thal. leucorrhoea.

Bulweria bulweri

July. 9. 10. "Terra Nova".
Procellaria pelagica.

Cardiff to South Trinidad

Heading South: June - July 1910

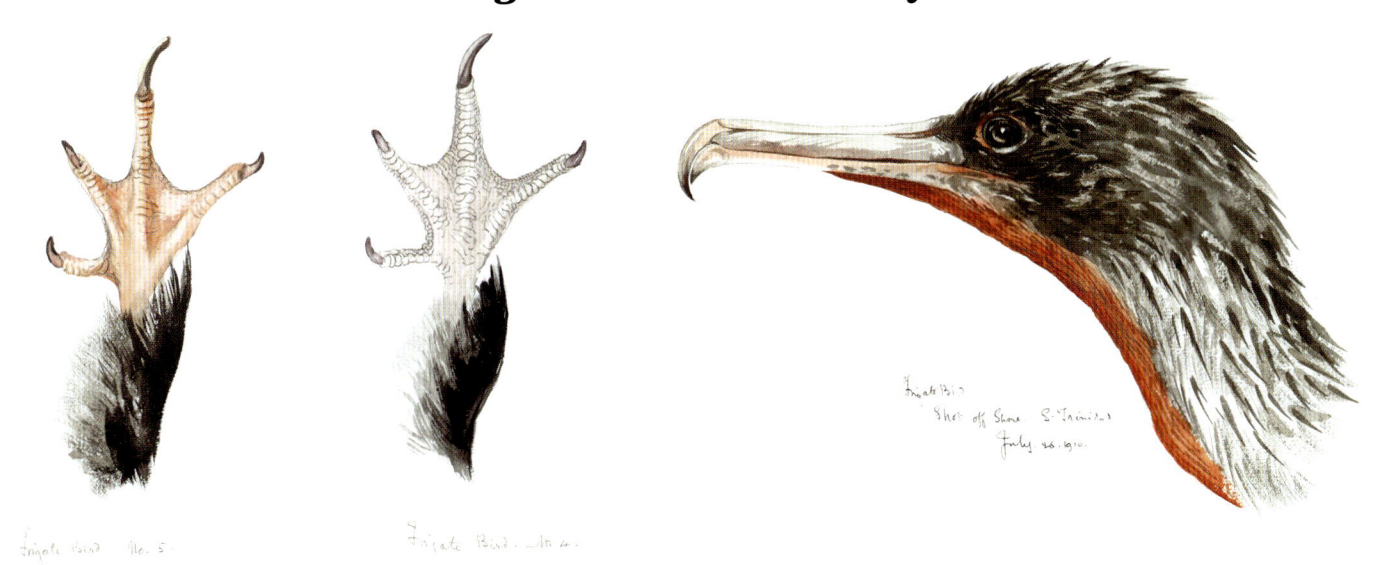

"Cherry Garrard and I shot a good series of birds, especially the Trinidad Petrel which has two phases, a black one and a light one. This bird has been given no less than three specific names but I am sure they all interbreed and are really the same species. Besides being called Oestrelata trinitatis it is also called Oe. arminjoniana and Oe. wilsoni after me - but we found every phase together nesting along cliff ledges, dark and pale - young and old." EAW

South Trinidad

The British Antarctic Expedition: 1910-1913

South Trinidad to Cape Town, South Africa

- 102 -

Heading South: July - October 1910

Cape Town, South Africa and Melbourne, Australia

"For my own part I have long been convinced that the first principle of right living is to put one's life into the hands of God and then do the work He gives one to do. I know one has sometimes to make a choice between two pieces of work which offer; but when one alone offers one is meant to do it trusting it will turn out all right in the end. One is at least not meant to shirk it because it is difficult or dangerous or disagreeable either to oneself or wife or parents or anyone else, and you would neither of you have me shirk this job." EAW

The British Antarctic Expedition: 1910-1913
R.M.S. Corinthic.

Cape Town, South Africa to Melbourne, Australia

Tuesday. Sept. 13.10. 38°57'S. 28°43'E. at noon.
Bill pink white. Tail black tip r. shaft.
D. regia. 1 all day.
D. exulans. Several all day, quite adult,
i.e. clean white backs, with sharp
cut black wings & a well marked
white patch on each wing.

One bird yesterday had clean white
back & quite black wings – no white
patch.
Several young ones today were brown
all over back & wings head & neck
on upper surface – whitish on the under
neck & throat except for broad brown
band under lower neck.

Young Diomedea exulans
can always be distinguished
from O.P. Ossifraga gigantea

by this – that D. ex. always at all ages
has white underwing with black tip
& edging – whereas all dark phases
of Oss. gig. have under wings dark, not
white. Otherwise, on the upper
parts both are very dark & show
no white.

Sept. 25.10. "Corinthic".
? Œst. bulleri.

Heading South: September - October 1910

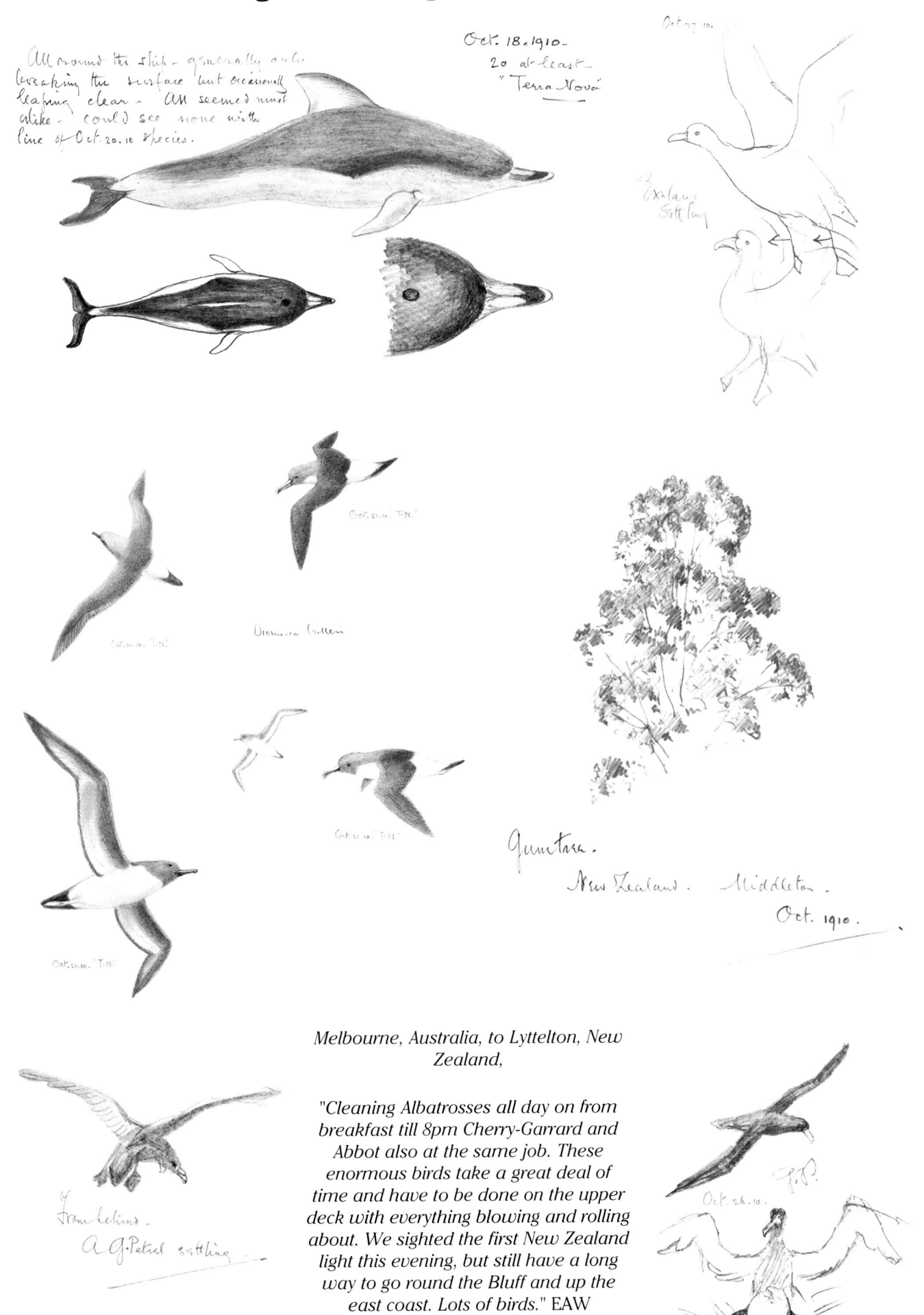

Melbourne, Australia, to Lyttelton, New Zealand,

"Cleaning Albatrosses all day on from breakfast till 8pm Cherry-Garrard and Abbot also at the same job. These enormous birds take a great deal of time and have to be done on the upper deck with everything blowing and rolling about. We sighted the first New Zealand light this evening, but still have a long way to go round the Bluff and up the east coast. Lots of birds." EAW

The British Antarctic Expedition: 1910-1913

Dec. 21.10. 10 pm.. Water sky. S.SW.
Cloudless weather - blowing to SE ✓

Deep blue (french)
to
Cobalt
v. pale
to
palest yell ochre white
to
Lilac or pure gray.

Cloud.. pure gray - Darker blue shadows below.
Light - ochreous.
Slanting shadows - dark bluish purple gray.
Sea white - ochreous faintly - with v dark gray line.
left half berg pure violet - right half pure blue -

Dec. 9. 10. 5 am - to E.

pure cobalt purple

black violet
pure cobalt

brownish
dull pale purple light -
dark inky shadows

Dec 9.10. Water washed bottom of overturned berg.

A Berg in the Pack. Ross Sea. Dec. 1910.

In the Pack Ice: December 1910

"It is difficult to say how the days go in the pack - one has so many odd things to do, such as examining distant seals, penguins, and birds with the glasses to say what they are, drawing icebergs at critical moments and various distances, putting in an hour now and then below to make a water colour sketch..." EAW

In the Pack Ice,
December 21 1910, 10pm

Pack Ice and Bergs in Ross Sea. December 1910

The British Antarctic Expedition: 1910-1913

Adélie Penguins in the Pack Ice, December 1910

In the Pack Ice: December 1910

"On the top of the berg sat about a hundred Antarctic Petrels like brown dots. They reminded me of the group of fleas one sees round the entrance of a Sand Martin's nest in the sunshine. The lower part of the berg, which had been burrowed into by the sea, was a wonderfully rich blue" EAW

The British Antarctic Expedition: 1910-1913

In the Pack Ice: December 1910

"The amount of life in the pack is its most astonishing feature - and the diatoms which support it all simply colour the ice red orange and yellow everywhere underneath. It is white above, of course - but the white of the snow is relieved everywhere with fissures and hollows of a wonderful cobalt and Prussian blue light, and along the edge washed by sea water the colour is peacock green." EAW

The British Antarctic Expedition: 1910-1913

In the Pack Ice: December 1910

"We saw a lot of the Adélie Penguins today, almost all young ones with the white throat, porpoising… They have lost none of their attractiveness - they are most comical and very interesting - as curious as ever, they will always come up at a trot when we sing to them, and you will often see a group of explorers on the poop singing 'for she's got bells on her fingers and rings on her toes, elephants to ride upon wherever she goes', and so on at the top of their voices to an admiring group of Adélie Penguins." EAW

The British Antarctic Expedition: 1910-1913

In the Pack Ice: January 1911

Pack Ice, Ross Sea, January 1911, midnight

"*Now and again one hears a penguin cry out in the stillness near at hand or far away, and then perhaps, he appears in his dress tail coat and white waistcoat suddenly upon an ice floe from the water - and catching sight of the ship runs curiously towards her, crying out in his amazement as he comes, from time to time, but only intensifying the wonderful stillness and beauty of the whole fairy-like scene as the golden glaring sun in the south just touches the horizon and begins again to gradually rise without having really set at all.*" EAW

Beaufort Island. Ross Sea, January 4 1911, 1 am

The British Antarctic Expedition: 1910-1913

Arrival at Cape Crozier: January 1911

"Love everything into which God has put life: and God made nothing dead. There is only less life in a stone than in a bud, and both have a life of their own, and both took life from God." EAW

The British Antarctic Expedition: 1910-1913

Cave in the Great Ice Barrier, Cape Crozier, January 4 1911

Arrival at Cape Evans and Depot Laying: January - March 1911

"We reached our final camp and made a large cairn and a depot of a ton of food for men, dogs and horses. It was very cold and windy, everyone getting frost nipped. The building up of this and the planting of a flag and 2 sledges and the sorting out of provisions took us all our time and we made no march. This is to be called One Ton Depot and it is in sight of the Bluff." EAW

The British Antarctic Expedition: 1910-1913

Arrival at Cape Evans and Depot Laying: January - March 1911

The rescue from the Sea Ice, March 1 1911

In the teeth of a blizzard

A crevasse rescue

"…. half way we met Atkinson who told us that they had now been joined by Scott and all the catastrophe party who were safe, but who had lost all the ponies except one - a great blow. However, no lives were lost and the sledge loads and stores were saved, so Meares and I returned to Hut Point to make stables for the only 2 horses that now remained..." EAW

The British Antarctic Expedition: 1910-1913

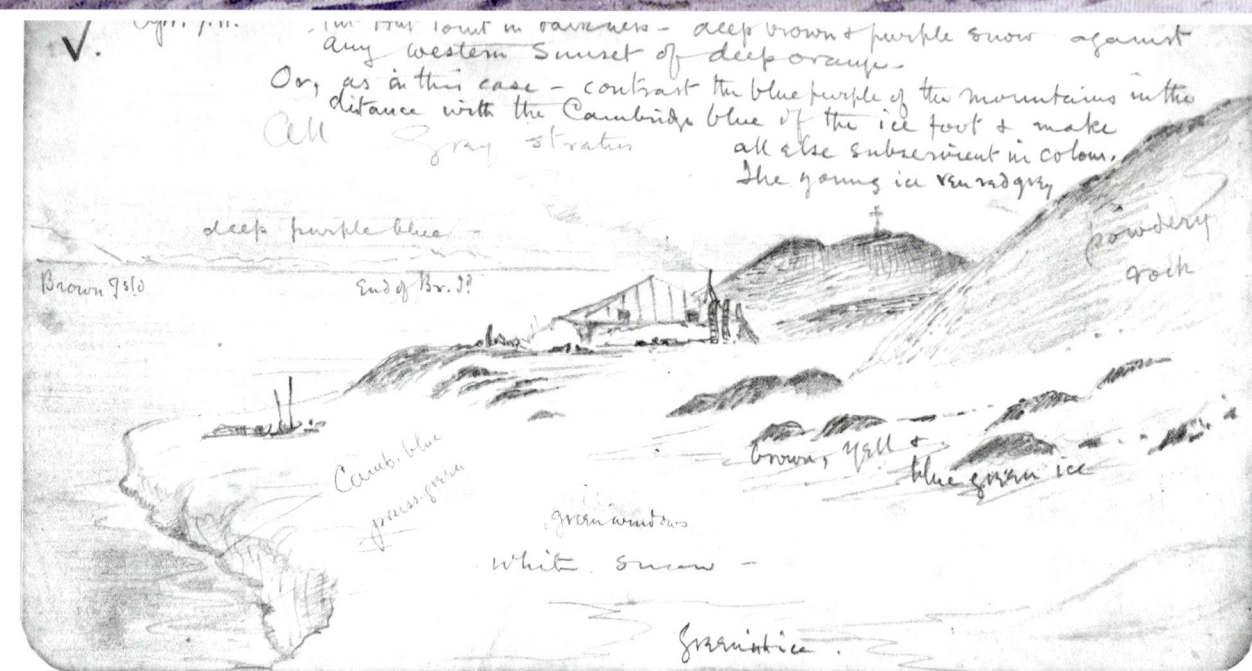

Waiting at Hut Point: March - April 1911

"Afternoon went to top of Observation Hill. Sketching at -7°F. in a very cold wind was difficult but the sunset was beautiful. Got some unusually good augite crystals out of an ash boulder. As we are getting the hut warmed the ice in the roof melts out and gives us a sort of snipe marsh to lie in at night" EAW

Hut Point - frostsmoke & new ice - Apr 3.11. 6 pm.

The British Antarctic Expedition: 1910-1913

Hut Point from the top of Observation Hill. April 1 1911, 5.30pm

Waiting at Hut Point: March - April 1911

"...I went up Observation Hill and sketched until it was dark. The lights were wonderful and the silence and stillness were absolute"
EAW

The British Antarctic Expedition: 1910-1913

Mount Erebus and Castle Rock from Observation Hill, March 31 1911

Waiting at Hut Point: March - April 1911

"The main thing is whether a person has the spirit of God in him, which to my mind means simply the power to love and be kind and unselfish; and many people have this in a very perfect form without professing any religious belief at all, or using any religious practices to keep it." EAW

Mount Erebus, April 28 1911

The British Antarctic Expedition: 1910-1913

Mount Erebus from Hut Point, March 1911

Waiting at Hut Point: March - April 1911

Sunset 7 p.m. March 30.11. Hut Point, Ski slope.

"*Out up Middle Crater forenoon and to top of Observation Hill afternoon. Grand sunset 5.30-7. Sketched up there comfortably for an hour. Wright and Taylor walked from Pram Point to the Gap on sea ice today.*" EAW

Looking across McMurdo Sound from Hut Point, March 29 1911, noon

The British Antarctic Expedition: 1910-1913

Mount Discovery, sunsets over new ice, March 1911

Waiting at Hut Point: March - April 1911

Observation Hill from Hut Point, April 9 1911, 6pm

"I have sketches enough to occupy me months when I get back to a paint box. This stay at Hut Point with so little to do has been a godsend to me at this time of the year when the colours are far the best, and most wonderful sunsets surprise one with some new light effect every day." EAW

McMurdo Sound, March 28 1911, 10pm

Observation Hill from Fodder Depot, April 14 1911, 2pm

The British Antarctic Expedition: 1910-1913

Frost Smoke, Looking North, Easter Sunday, April 16 1911

Sunset, Looking West, from Hut Point, April 1 1911

Waiting at Hut Point: March - April 1911

"It's not what happens to you, it's what you make of it that matters." EAW

The British Antarctic Expedition: 1910-1913

Looking N. from Hut Point. Very young ice & frost smoke. Apr. 2.11. 7 pm.

Frost Smoke and new ice in McMurdo Sound, March 30 1911, 7pm

Return to Cape Evans: April 1911

Down the Hutton Cliffs, April 21 1911

Sunset, McMurdo Sound. April.13.11. 5pm.

"We had a lot of heavy pulling going over the hills going along the top of the promontory to Hutton Cliffs and here we all went down a 20ft ice cliff by rope on to the sea ice in a smother of snow as a small blizzard came on. We then also lowered the loaded sledges one by one, camped under a cliff, and had some tea 4pm." EAW

April.29.11. Noon, looking South.

The British Antarctic Expedition: 1910-1913

Berg off Cape Evans. April 23 1911, Last day of the sun

Cape Barne from Cape Evans. April. 28. 11. Noon.

Last of the Sun - Cape Evans: April 1911

Barne Glacier from the Ramp, April 30 1911, 1pm

"After dinner I often spend the evenings painting odd sketches made round our winter quarters and my attempts at the splendid afterglows and sunsets and twilight colouring which we get so much of now, all being done at the ward-room table under the public eye, afford a good deal of amusement." EAW

Frost smoke at Sunset, Easter Sunday, April 16 1911

The British Antarctic Expedition: 1910-1913

Paraselena, Cape Evans, May 13 1911, 8am

Antarctic Winter - Cape Evans: 1911

"Atkinson is using the fish trap and caught 40 notothenia today in the morning and 41 in the evening - at least 3 species, but all very much alike, not as much flesh on 2 of them as there is on a Whiting. I painted 2 of them - one reddish orange the other bluish brown. They are full of parasites, protozoa, nematodes, trematodes and a copepod on the gill covers. We eat them all the same, but I can't say they are nice." EAW

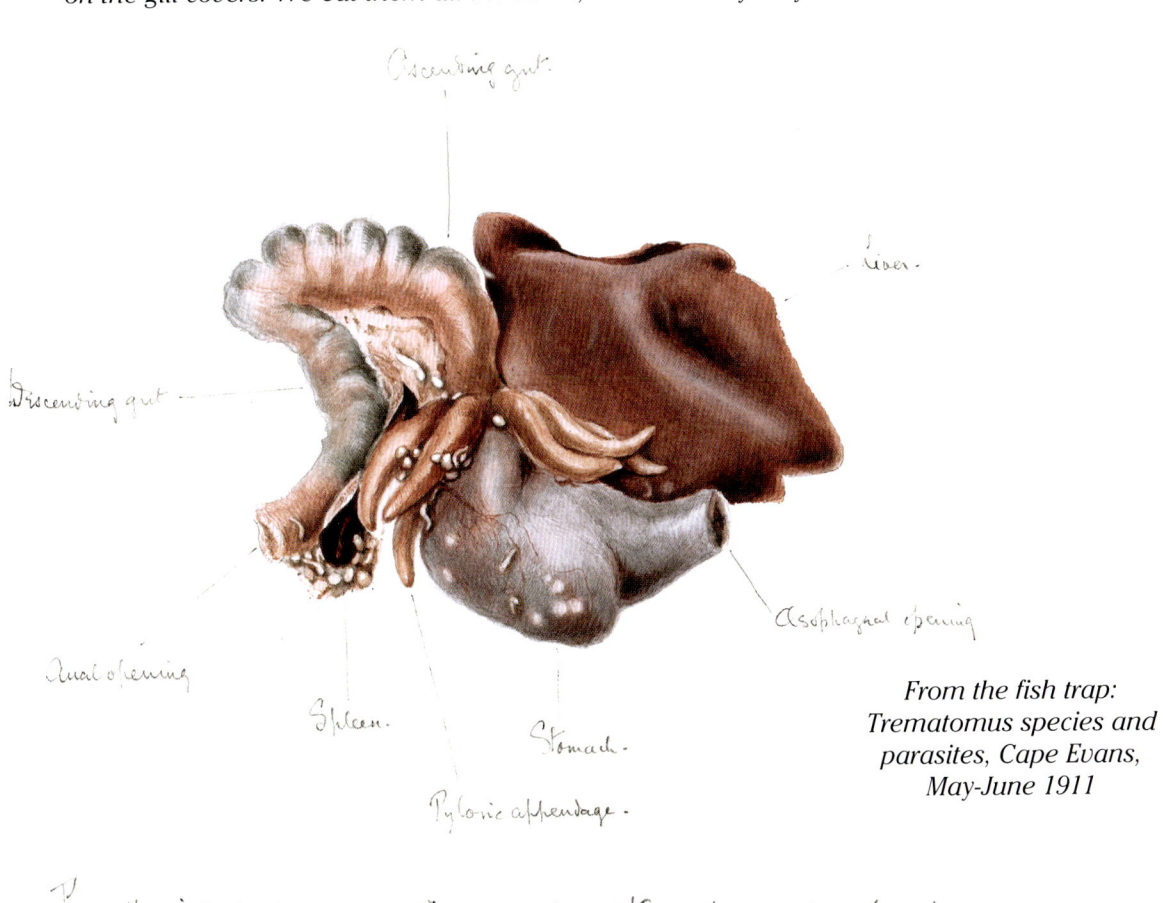

From the fish trap: Trematomus species and parasites, Cape Evans, May-June 1911

The British Antarctic Expedition: 1910-1913

Birdie Bowers, reading the thermometer on the Ramp, June 6 1911

Antarctic Winter - Cape Evans: 1911

"*Again had a blizzard headache and the blizzard came right enough with gusts up to 72 miles an hour. Hurricane force is 75 for an hour. I was nearly blown off the roof when clearing the anemometer. Gave the day up to preparing my lecture for tonight on sketching and this came off all right.*" EAW

Mt. Erebus at noon, May 28 1911

The British Antarctic Expedition: 1910-1913

MENU FOR MIDWINTER DAY 1911.   **CAPE EVANS McMURDO SOUND**

BRITISH ANTARCTIC EXPEDITION — TERRA NOVA R.Y.S.

- CONSOMMÉ · SEAL
- ROAST BEEF & YORKSHIRE PUDDING
- HORSE RADISH SAUCE
- POTATOES A LA MODE & BRUSSELS SPROUTS
- PLUM PUDDING · MINCE PIES
- CAVIARE ANTARCTIC
- CRYSTALLISED FRUITS · CHOCOLATE BONBONS
- BUTTER BONBONS · WALNUT TOFFEE
- ALMONDS & RAISINS
- WINES
- SHERRY · CHAMPAGNE · BRANDY PUNCH · LIQUEUR
- CIGARS · CIGARETTES & TOBACCO
- SNAPDRAGON
- PINE-APPLE CUSTARD · RASPBERRY JELLIES
- BUSZARD'S CAKE

GOD SAVE THE KING.

Mid-Winter - Cape Evans: June 1911

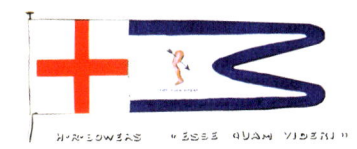

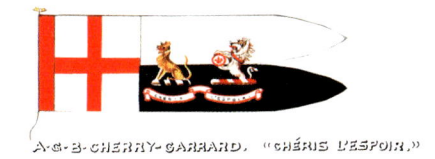

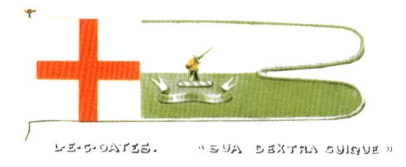

H.R. BOWERS "ESSE QUAM VIDERI" A.G.B. CHERRY-GARRARD "CHÉRIS L'ESPOIR" L.E.G. OATES "SUA DEXTRA CUIQUE"

"The forenoon I spent painting a large menu for signatures which is to appear in the S.P.T. 2nd part of Vol. III. At lunch we had the hut decorated with all our sledge flags and a huge cake and then the S.P.T. first part was handed by the editor to the Owner who read it out loud amidst much amusement." EAW

Captain Scott

- 143 -

The British Antarctic Expedition: 1910-1913

Erebus Slopes and the Ramp from Cape Evans, June 11 6pm

The Winter Journey to Cape Crozier: 27 June -1 August 1911

"Sixth Sunday after Trinity and quite the funniest birthday I have ever spent. The wind was terrific. It blew almost continuously with storm force - there were slight lulls occasionally followed by squalls of very great violence, and at about noon the canvas roof of the hut carried away and we were left lying exposed in our sleeping bags without a tent or a roof" EAW

THE SOUTH POLAR TIMES   80

1. THIS is the House that Cherry built.[45]

2. THIS is the Ridge that topped the Moraine
   That supported the House that Cherry built.

3. THESE are the Rocks and Boulders "Erratic",
Composing the walls - with lavas "Basic" -
   That stood on the Ridge that topped the Moraine etc.

4. THIS is the Sledge and Canvas strong
That formed a roof about ten feet long,
   To cover the Rocks and Boulders "Erratic",
Composing the Walls - with lavas "Basic" - etc.

5. THESE are the Ice-blocks hard and stout
That were placed so carefully round about,

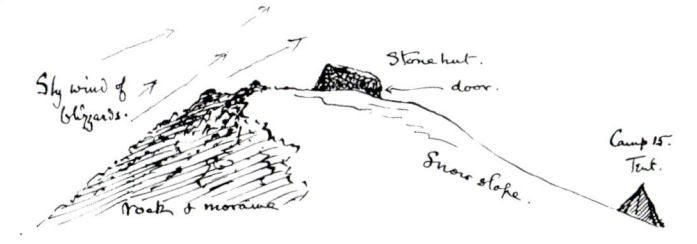

- 145 -

The British Antarctic Expedition: 1910-1913

The Return of the Sun: August - September 1911

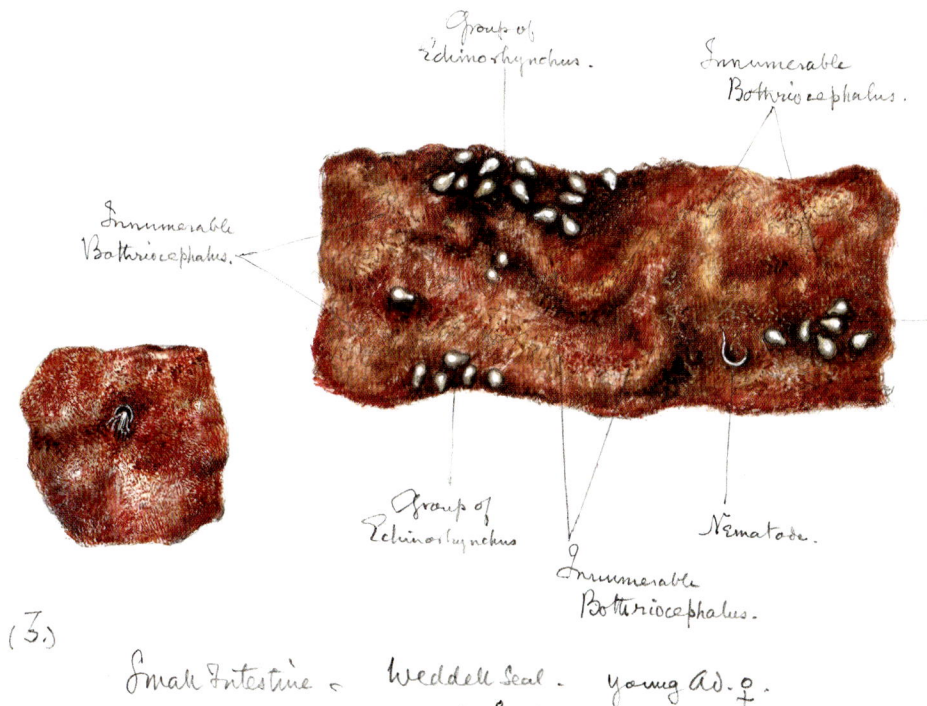

"Had to start exercising my pony today - taking him out for 2 hours a day. My pony is Nobby, one of the 2 survivors of last season's depot journey. He is slow, but one of the quietest and most reliable of the lot" EAW

Group of Echinorhynchus.
Innumerable Bothriocephalus.
Innumerable Bothriocephalus.
Group of Echinorhynchus.
Group of Echinorhynchus.
Nematode.
Innumerable Bothriocephalus.

Emperor Penguin.
(Aptenodytes forsteri)
Aug. 16. 1911.

(3.) Small Intestine - Weddell Seal. young ad. ♀.
(Leptonychotes weddelli)
Aug. 9. 11.

(a.) Peritoneal surface of main intestine, showing cysts & adhesions.

(b.) Mucous surface of the same, showing entrance to one of the cysts & a bunch of Cestodes. Villi engorged uniformly with blood.

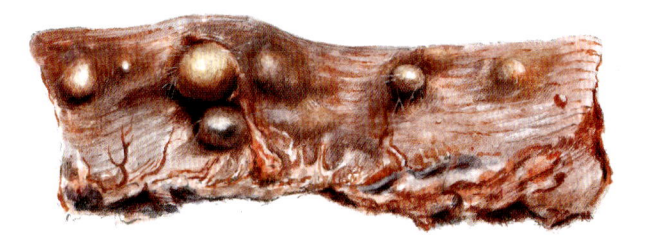

The British Antarctic Expedition: 1910-1913

Iridescent clouds, looking north from Cape Evans, August 9 1911

The Return of the Sun: August - September 1911

Mt Erebus, showing drift of smoke, August 13 1911, 10am

"*Calm clear day, perfect weather. Colour everywhere. Went up the Harbour Hills. Erebus' smoke was lit up with a fiery orange light by the set sun. The blue on the southern and western horizon rising into lilac and rose pink was exactly what one sees in the Swiss winter. Here, now that the sun is below the horizon, one sees the shadow of the slopes of Erebus thrown against the sky in a pure blue on the south-west horizon. Afternoon drawing, evening painting by the acetylene.*" EAW

Spiral Cirrostratus, August 11 1911

The British Antarctic Expedition: 1910-1913

Inaccessible Island and Bergs, looking S. from Barne Glacier, August 13 1911

The Return of the Sun: August - September 1911

"Heaven is with us here, not in the skies to be reached only after death… Whenever we see God in another we get a glimpse of the place where He is and that is the joy of heaven." EAW

The British Antarctic Expedition: 1910-1913

The Return of the Sun: August - September 1911

"Each child of God, each one of us, has a separate part to play in upholding the perfection of God's Wisdom, and of bringing it to perfection in ourselves; some by fasting and an ascetic life, some by a convivial and social life." EAW

The British Antarctic Expedition: 1910-1913

Parhelion from the Ramp, September 14 1911

A Prismatic Parhelion, September 11 1902

The Return of the Sun: August - September 1911

"Painting all remainder of day and began doing circular pictures, a shape that suits these sketches that one gets down here with so much horizon and horizontal skies and floes."
EAW

Mount Erebus, September 1911

The British Antarctic Expedition: 1910-1913

The Return of the Sun: August - September 1911

Iceberg studies

"This just about finishes the S.P.T. for this third volume, and it is to be published on Sunday. It is a very poetical number - or rather versical - I am not sure that there is much poetry in any of the verses. I have even run to two verses myself. Pathological I'm afraid - perhaps an early symptom of polar anaemia." EAW

The British Antarctic Expedition: 1910-1913

Field sketches, crossing the Great Ice Barrier, November 1911

The Pole Journey: November 1911 - March 1912

Field sketches, ascending the Beardmore Glacier, December 1911

"Magnificent ochreous reddish gneissic granite columnar crags and pillared mountains on both sides, the western end of Mt. Hope being low and rugged with the same nearly all in situ - not much scree anywhere and what angle the valley sides make is always about 45°. Enormous cwms on our right low down and full of snow with avalanches here and there and in one or two places the screes ended at the top in a sort of sheared off table-land of rock. The mountains of gneissic granite look almost like columnar basalt. They are weathering out vertically and their tops are flat at about 3,000 ft with a dip to the west…" EAW

The British Antarctic Expedition: 1910-1913

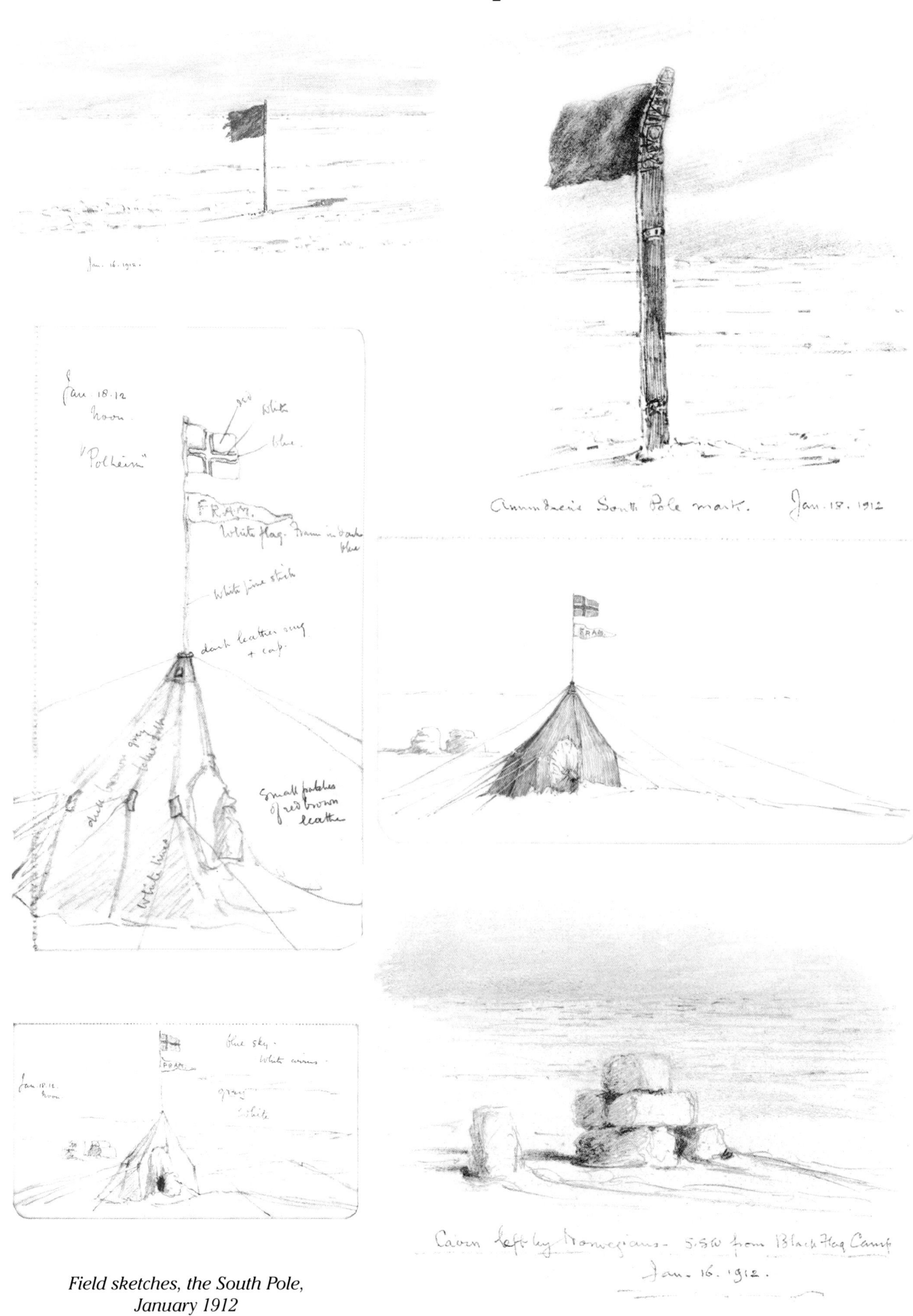

*Field sketches, the South Pole,
January 1912*

The Pole Journey: November 1911 - March 1912

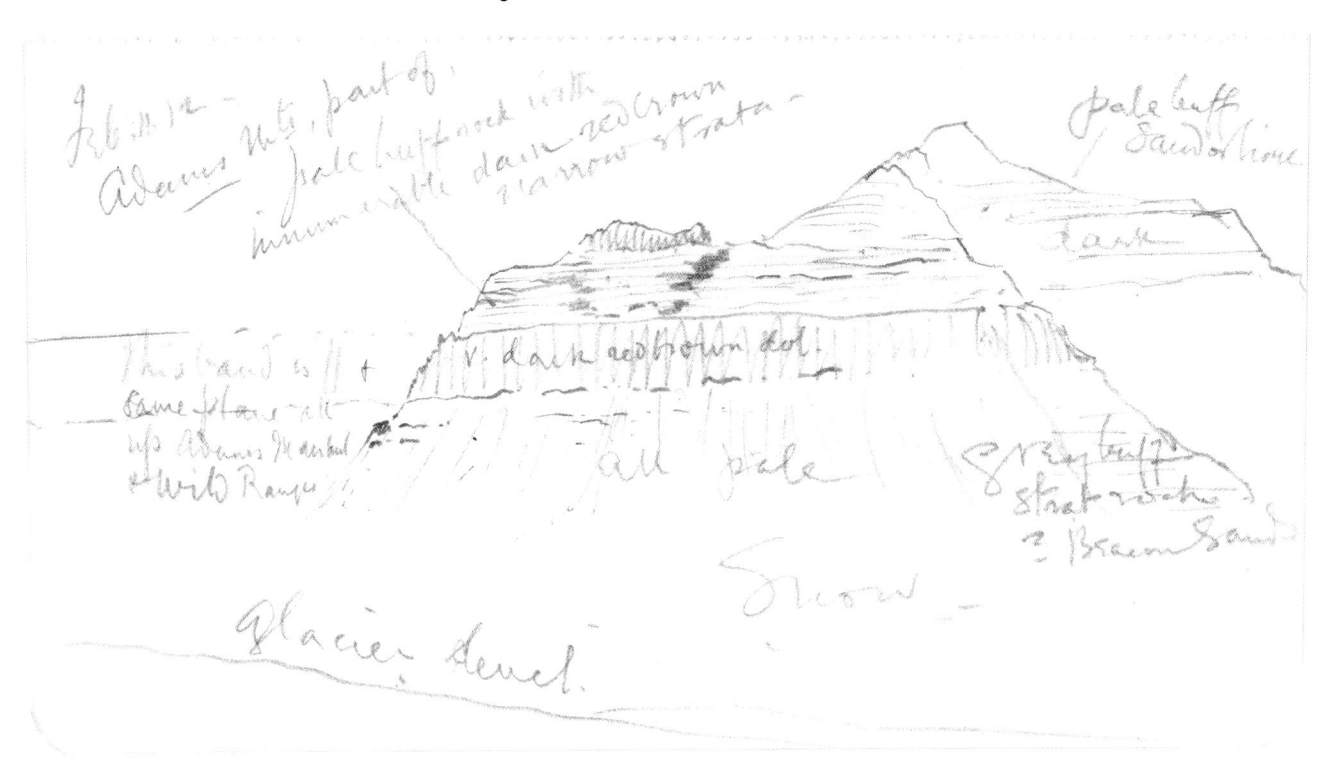

Field sketches, the Polar Plateau and descent of the Beardmore Glacier, February 1912

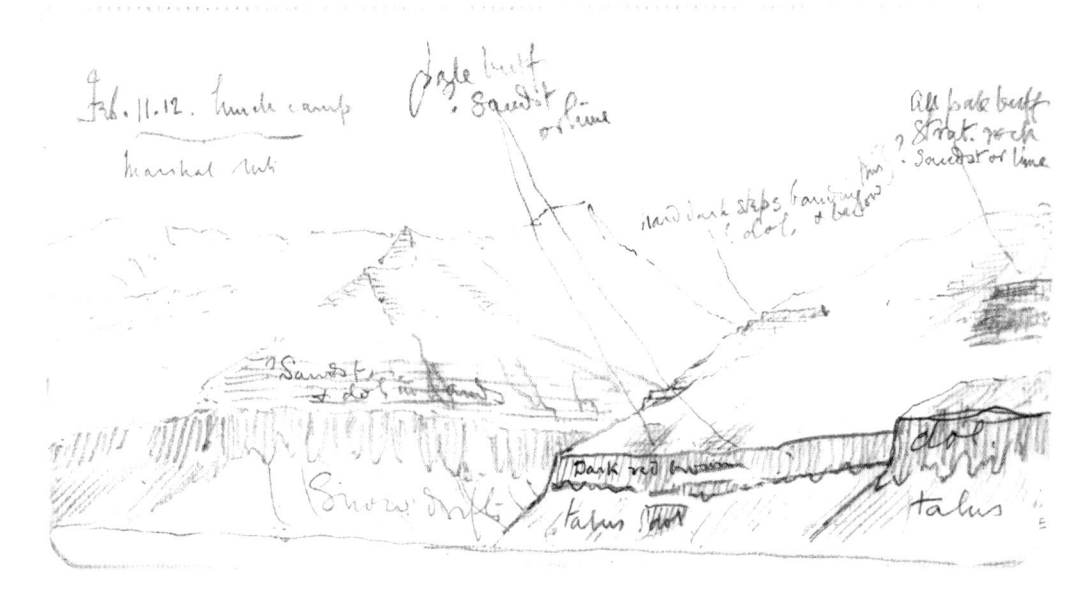

"*After lunch we all geologised until supper, and I was very late turning in, examining the moraine after supper. Socks all strewn over the rocks dried splendidly. Magnificent Beacon Sandstone cliffs. Masses of limestone in the moraine - and dolerite crags in various places. Coal seams at all heights in the sandstone cliffs and lumps of weathered coal and fossils, vegetable. Had a regular field day and got some splendid things in the short time.*" EAW

In Memoriam - One Hundred Years On: 1912-2012

The deaths of Captain Scott and the Pole Party started an international media frenzy which helped to build a myth. The courage of Scott and his companions moved a generation. Across the nation and the Empire, memorial services were held. More than ten thousand people stood outside St. Paul's Cathedral unable to get in for the national memorial service, led by the King. Posthumous awards and medals were bestowed upon the Pole Party. Scott's final written appeal to the country to "look after our people" resulted in a huge response which raised £75,000 (equivalent to around £4,500,000 today). From this, all of the dependants of the men who died received pensions.

For all that Ted would have disliked it, he became catapulted into the national arena as a hero of the British Empire. For many years his name was widely recognised and he appeared on cigarette cards, postcards and tea caddies. A statue of Ted was created by Lady Scott and unveiled on 9 July 1914 by Sir Clements Markham in front of a large crowd on the Cheltenham Promenade. Ted had also been a popular figure amongst his friends and colleagues and the fond remembrances which they had of him resulted in numerous other monuments. Memorial windows to Ted were installed at Cheltenham College and Copthorne School, Sussex; Ted's sledging flag was hung in the Lady Chapel of Gloucester Cathedral (it has since been moved to the Scott Polar Research Institute); a Caius College flag which Ted took to the Pole hangs in the dining room of Gonville and Caius College to this day; a memorial was unveiled at Cortachy in Scotland, placed by the Reginald Smiths; and an Edward Wilson room was installed at St. George's Hospital. More recently social housing projects in Cheltenham have been named for him, and a primary school in London, close to where he used to work in the Caius Mission, with the children from the slums of Victorian Battersea.

Frigata ariel wilsoni, a type subspecies of Frigatebird collected by Edward Wilson from South Trinidad

Medals and statues would have flattered him, overwhelmed him even, but not impressed him. In many ways, important though they are, they only partly reflect the man. Over the past hundred years as those who remembered him have died and the events which surrounded his life have faded into history, fewer people remember his life and works. Today, Ted's legacies are, perhaps, more like his life, subtle and complex, often unnoticed in the background, and deeply interwoven, each with the other.

His artistic legacy is a considerable one. A retrospective exhibition of Ted's pictures was held at the Alpine Club in London in December 1913 and was visited by Queen Alexandra and other members of the Royal Family. Since then small exhibitions of his paintings have irregularly taken place and are always well attended. However, few of his pictures are publicly accessible and his artistic achievements are generally under-appreciated as a result.

When George Seaver published his 3 volumes of biography in the 1930's the example of Ted's life itself became an inspiration to many. Perhaps more so after the 1948 film, *Scott of the Antarctic* appeared in cinemas world-wide. The extent and depth of Ted's spiritual life was known only to his closest family and friends when he was alive and Seaver's books were a revelation even to many of his comrades on the expeditions. His ideas were not, in many ways, original but Ted was a seeker of Truth not originality: he sought to turn the teaching of the Christian Gospel and traditions (especially monastic traditions) into the reality of his ordinary every day life. The Faith of Edward Wilson was taught in many schools as a part of Religious Education classes and in South Africa, for a time, was a compulsory part of the curriculum. The Scouts even had a 'senior scout' patrol badge named for him. His life and faith inspired many others to become doctors, missionaries, scientists, artists and priests. However we live in an increasingly secular age and his faith is as likely to be a source of derision a century on, as of admiration. His life exemplar does not sit well in an age of cynicism and self-indulgent excess. Nevertheless, Ted continues to inspire the lives of many, if not on the widespread scale once commonplace.

Finally, perhaps, there is his scientific legacy. Many fields of Antarctic science have been invisibly touched by him, if only through his work in encouraging others, particularly his scientific staff on the *Terra Nova* Expedition. Our knowledge of the Antarctic today owes much to his work and inspiration. There are reams of physiographical and geological notes and drawings in addition to his own biological work.

Edward Wilson's Christening mug

The Winter Journey was one of the most astonishing journeys to be successfully executed throughout the Heroic Age of Antarctic exploration. The scientific hypotheses on which the journey was based subsequently turned out to be false but it still stands as one of the most extraordinary of scientific quests. It might not have proved the missing link between dinosaurs and birds that was looked for but for Ted, it was the search for Truth that was important, not worldly success. However, he would no doubt have been delighted that a century later the eggs are the most asked after aspect of the ornithological collections in the Natural History Museum, such is the interest in their story. The specimens also now represent priceless samples from a pre-pollution Antarctic and are of considerable scientific value. Many of his other ornithological specimens are also of unique importance. In another field altogether, however, Ted did provide the missing link. The specimen of *Glossopteris* from the Beardmore Glacier proved to be crucial evidence that the Southern Continents were once joined, helping to provide evidence for the Gondwana theory of a Southern super-continent, leading into the Theory of Continental Drift. The specimens which were carried with the Pole Party to the day they died, at Ted's specific request, changed the way in which we understand the world.

The specimen of Glossopteris collected by Edward Wilson from the Beardmore Glacier

His scientific legacy is, however, even more considerable than this. The scientific work undertaken under Ted's direction during the last expedition inspired all involved. It meant that when there was money left over from the appeal for the widows and orphans, it was used to help found a Scott Polar Research Institute in Cambridge as a national memorial to Scott and the Polar party. The Institute is as much a credit to his Chief of Scientific Staff as it is to Scott. It carries on the multi-disciplinary work of Scott's expeditions to this day, continuing the extraordinary legacy of the expedition in tackling the important scientific issues of the day and ever increasing our understanding of the Polar Regions. In addition, the Institute holds many of Ted's Antarctic pictures. It also administers two funds in his memory. The first gives small grants to encourage science and painting in the Polar Regions, the second helps to administer the polar art collections of the Institute. All the royalties from this book will be going to help endow these memorial funds.

Yet Ted's work extended beyond the Antarctic. In some small way we still have penguins to enjoy because of Ted's move towards conservation campaigning. In the United Kingdom we still have grouse, in part due to his work. And perhaps this is the biggest legacy of all of Ted's achievements - the story of love, loyalty and friendship: human care in its highest form both for his friends and for the planet.

Scott could have written many rather obvious things to his wife about the future of their son, Peter, as he lay dying on the Great Ice Barrier in March 1912: "send the boy into the Royal Navy"; "make the boy a great explorer"; or "send him to Eton" seem obvious choices. Instead, inspired by the friend who lay dying by his side, Scott wrote "Make the boy interested in natural history if you can; it is better than games; they encourage it at some schools." Lines that perfectly echo the strongly held views upon the subject, which Ted had held since his days at Cheltenham College. From the deep friendship of two men, dying together on the Great Ice Barrier, sprang the inspiration that would forge the life of Sir Peter Scott, who grew up to become another of the 20th century's great water-colour painters, naturalists, ornithologists and a leading conservationist. Sir Peter was not any old conservationist, however. David Bellamy described him as 'the father of conservation' and Sir David Attenborough called him the 'patron saint'. He became the founder of the Wildfowl and Wetlands Trust and of the World Wildlife Fund (World Wide Fund for Nature) and many other world wide conservation institutions and movements. His work has forged important aspects of our modern lives through the conservation movement, even down to the governance of the Antarctic through its Environmental Protocol.

The life of Edward Wilson was taught in many schools and youth groups. This Scout Patrol Badge was used for Senior Scout 'Wilson Patrols' from c.1960-67

Whether we know it or not, through these many astonishing legacies, Ted, has touched the lives of us all. As he appears on centenary stamps, pins and other centenary souvenirs to mark the Scott 100, it is right that we should recall his often forgotten legacy to us all. Not least amongst these are the thousands of watercolours and sketches which he has left us, a few of which we reproduce here. We hope that you have enjoyed them.

In Memoriam - One Hundred Years On: 1912-2012

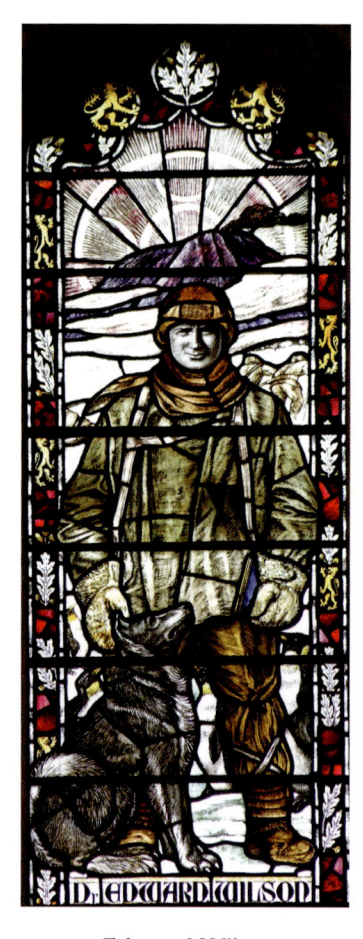

Edward Wilson Memorial Window, Cheltenham College

Edward Wilson Statue, Cheltenham

Edward Wilson Memorial Window, Copthorne School

Unveiling the statue of Edward Wilson, Cheltenham, 1914

In Memoriam - One Hundred Years On: 1912-2012

A balaclava which belonged to Edward Wilson. Items which belonged to the great explorer are highly sort after by Museums and Private Collectors alike

Artefacts related to Edward Wilson on display at Cheltenham Art Gallery and Museum

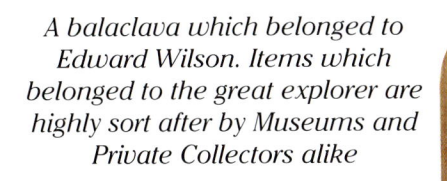

Although rarely on show, a century on, the Emperor Penguin eggs collected by Edward Wilson during the Worst Journey in the World are the most asked after items in the ornithological collections of the Natural History Museum

Many art collections contain works by Edward Wilson, including that of Her Majesty, Queen Elizabeth II

The final camp

The Polar Museum in Cambridge displays many artefacts relating to Edward Wilson, including paintings, memorabilia and last letters

In Memoriam - One Hundred Years On: 1912-2012

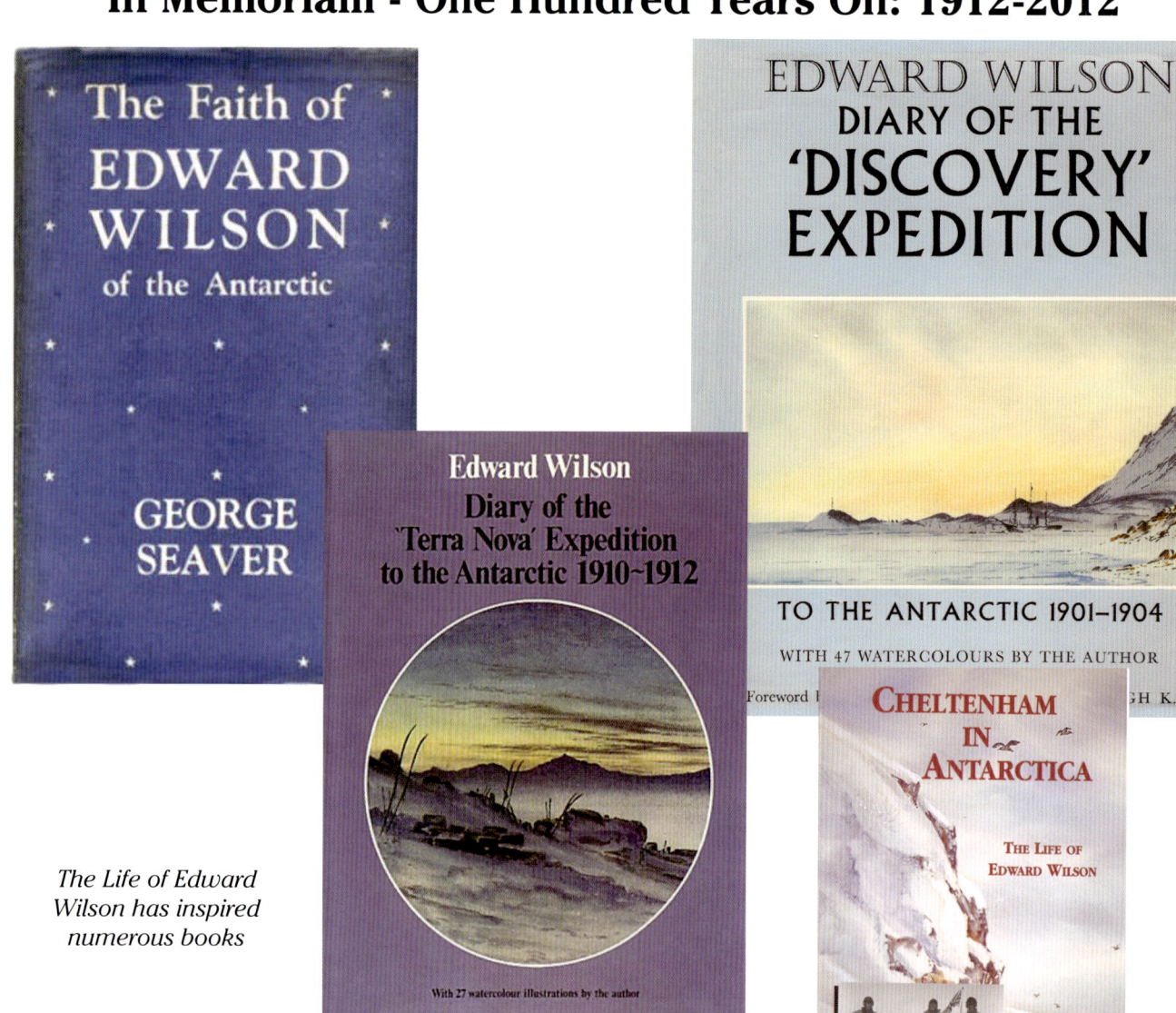

The Life of Edward Wilson has inspired numerous books

Edward Wilson has several buildings named in his memory. The Edward Wilson Primary School is in London, near to where he worked in the Battersea Slums

The Life of Edward Wilson has also inspired poetry, plays and novels. Harold Warrender played Edward Wilson in the famous 1948 film, Scott of the Antarctic, with its famous score by Vaughan Williams

In Memoriam - One Hundred Years On: 1912-2012

Edward Wilson's life and paintings have appeared on all sorts of collector's souvenirs over the past century, from postcards and prints, to tea-caddies, stamps, cigarette cards and pins…

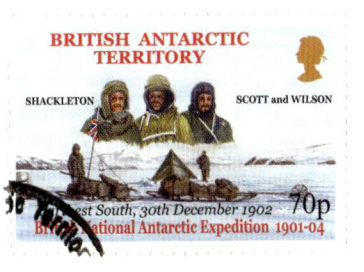

- 167 -

Select Bibliography and Further Recommended Reading

Armitage, A.B.	*Two Years in the Antarctic: Being a Narrative of the British National Antarctic Expedition.* London, Edward Arnold, 1905
Barrett Hamilton, G.	*A History of British Mammals.* Parts 1-21. Illustrated by & Hinton, M.A. E.A.Wilson & Guy Dollman. London, Gurney & Jackson, 1910-1921
Chapman, A.	*Bird-life of the Borders: Records of Wild Sport and Natural History on Moorland and Sea.* 2nd edition. London, Gurney & Jackson, 1907 [Frontispiece by E.A.Wilson]
Cherry Garrard, A.	(ed.) *The South Polar Times,* Vol. III. London, Smith Elder & Co., 1914 *The Worst Journey in the World.* London, Constable, 1922
Fox, W.	*Terra Antarctica: Looking Into the Emptiest Continent.* San Antonio, Texas, Trinity University Press, 2005
Higgins, H.	*The Semilunar Fibro-cartilages and Transverse Ligament of the Knee Joint,* illustrated by E.A.Wilson. London, *Journal of Anatomy and Physiology* Vol. 29, 1895
King, H.G.R.	(ed.) *Edward Wilson: Diary of the 'Terra Nova' Expedition to the Antarctic 1910-1912.* London, Blandford Press, 1972 (ed.) *South Pole Odyssey: Selections from the Antarctic Diaries of Edward Wilson.* Poole, Blandford Press, 1982
Knipe, H.R.	*Nebula to Man.* London, Dent, 1905 [part illustrated by E.A. Wilson]
Lack, D.L.	*Some British pioneers in Ornithological Research 1859-1939.* London, *Ibis*, Vol. 101 #1, 1959
Leslie, A.S. & Shipley, A.E.	(eds.) *The Grouse in Health and in Disease: Being the Final Report of the Committee of Inquiry on Grouse Disease.* London, Smith Elder & Co., 1911
Prichard, H.	*Sport in Wildest Britain.* "Illustrated from water-colour paintings by Dr. Edward A. Wilson". London, William Heinemann, 1921
Roberts, B.	(ed.) *Edward Wilson: Birds of the Antarctic.* London, Blandford Press, 1967
Rolleston, H.D.	*Diseases of the Liver, Gall Bladder and Bile-Ducts.* Illustrated by E.A.Wilson. Philadelphia, W. B. Saunders, 1905
Savours, A.	(ed.) *Edward Wilson: Diary of the Discovery Expedition to the Antarctic Regions 1901-1904.* London, Blandford Press, 1966
Scott, R.F.	*The Voyage of the Discovery.* London, Smith Elder & Co., 1905 *Scott's Last Expedition,* London, Smith Elder & Co., 1913
Seaver, G. London,	*Edward Wilson of the Antarctic: Naturalist and Friend.* John Murray, 1933 *Edward Wilson Nature Lover.* London, John Murray, 1937 *The Faith of Edward Wilson.* London, John Murray, 1948
Shackleton, E.H. & Bernacchi, L.C.	(eds.) *The South Polar Times,* Vols. I & II. London, Smith Elder & Co., 1907
Skelton, J.V. & Wilson, D.M.	*Discovery Illustrated: Pictures from Captain Scott's First Antarctic Expedition.* Cheltenham, Reardon Publishing, 2001
Solomon, S.	*The Coldest March: Scott's Fatal Antarctic Expedition.* New Haven, Yale University Press, 2001

Walker, C.E.	*Old Flies in New Dresses: How to Dress dry Flies with the Wings in the Natural Position and Some New Wet Flies.* Illustrated by E.A. Wilson & the author. London, Lawrence & Bullen, 1898
Williams, I.	*With Scott in the Antarctic: Edward Wilson Explorer, Naturalist, Artist.* Stroud, The History Press, 2008
Wilson D.M. & Elder D.B.	*Cheltenham in Antarctica: The Life of Edward Wilson.* Cheltenham, Reardon Publishing, 2000
Wilson, D.M. & Wilson, C.J.	*Edward Wilson's Nature Notebooks.* Cheltenham, Reardon Publishing, 2004
Wilson, E.A.	*Notes on Antarctic Seals collected during the Expedition of the 'Southern Cross'* in *Report on the Collections of Natural History made in the Antarctic Regions during the Voyage of the 'Southern Cross'.* London, British Museum (Natural History), 1901 *The Birds of the Island of South Trinidad.* London, *Ibis.* 8th series, Vol.4 #14, 1904 *On Some Antarctic Birds.* London, *Proceedings of the IVth International Ornithological Congress,* 1905 *Exhibit and Discussion on Albino Penguins* (with W.E.Clarke, & Lord Rothschild). London, *Bulletin of the British Ornithologist's Club,* Vol.15 #114, 1905 *International Bird Protection.* London, *Bird Notes and News,* No.10, July 1905 *Penguins.* London, *Bird Notes and News,* No.11, October 1905 *The Distribution of Antarctic Seals and Birds.* London, *Geographical Journal,* Vol. 25 #4, 1905 *The Emperor Penguin.* London, *Ibis.* 8th series, Vol.5 #18, 1905 [Summary of a lecture given at the Royal Institution on 27 January 1905. Reprinted from *The Times,* 28 January 1905] *Aves,* in *National Antarctic Expedition 1901-1904, Natural History Vol. II Zoology.* London, British Museum (Natural History), 1907 *Mammalia,* in *National Antarctic Expedition 1901-1904, Natural History Vol. II Zoology.* London, British Museum (Natural History), 1907 *National Antarctic Expedition 1901-1904: Album of Photographs and Sketches with a Portfolio of Panoramic Views.* London, Royal Society, 1908 *The Changes of Plumage in the Red Grouse (<u>Lagopus scoticus</u>) in Health and in Disease.* London, *Proceedings of the Zoological Society of London,* December 1910 [n.b. the ZSL proceedings for 1909/1910 contain articles by Shipley, Fantham and others involved with the Grouse Disease Inquiry which are illustrated by E.A.Wilson] *The British Antarctic Expedition.* London, *Geographical Journal,* Vol 39 #6, 1912
Yelverton, D.E.	*Antarctica Unveiled, Scott's First Expedition and the Quest for the Unknown Continent.* Denver, University Press of Colorado, 2000

There are a large number of publications relating to Scott's two Antarctic Expeditions containing illustrations by E.A.Wilson and references to his life and work. These are, however, too numerous to provide a comprehensive listing.

List of Illustrations and Copyright Acknowledgements

With thanks to:

SPRI = Scott Polar Research Institute
CAGM = Cheltenham Art Gallery and Museum
NMBL = National Marine Biological Library, Plymouth
BAL = Private Collection. Photo © Christie's Images/The Bridgeman Art Library
Cheffins = Cheffins Fine Art Auctioneers, Cambridge
NHM = Natural History Museum, London
AHAG = Abbot Hall Art Gallery and Museum, Kendall
DHT = Dundee Heritage Trust
RGS = Royal Geographical Society
CC = Cheltenham College
The Royal Collection = Her Majesty Queen Elizabeth II
Private = Numerous Private Collections

****

Endpaper

Ice Island Berg
and Pack ice from the Crow's Nest,
Ross Sea. 19 Dec. 1910 6.30am © SPRI N:456

Frontispieces

p01   - Edward A. Wilson signature. Undated © private
      - Notebook stickers. Undated © SPRI Wilson Polar Sketchbook 1801
      - McMurdo Sound from Arrival Heights. 29 Aug. 1911 © SPRI N:511

p02   - Adélie Penguin heads, for *Discovery* Expedition Scientific Reports. 1907 © SPRI Y:67/5/13

p03   - Sea Leopard [Leopard Seal] Chasing Emperor Penguins. Sketch for *South Polar Times.* May 1902 © SPRI N:1426

p04   - Caught in a Blizzard. Undated © BAL CH375980

p05   - Emperor Penguin chicks reading the *South Polar Times*. c.1905 © CAGM
      - A sudden turn. Armitage 1905 © private
      - Half sleeping bags and blouses. Armitage 1905 © private

P06   - Please Open the Door at Once. Armitage 1905 © private
      - Emperor Penguin. Armitage 1905 © private

p07   - Icebergs. 1901-1904 © SPRI Y:61/7

The Formative Years: 1872-1901

p08   - Stanmore Common. c.1900 © SPRI Y:2006/15/24a

p09   - Peewit [Lapwing] sketches, Fishmonger's bird. Jan. 1900 © private
      - Peewit [Lapwing] study for wing. Jan. 1900 © private
      - Golden Plover, winter. Plumage studies. Jan. 1900 © private
      - Narcissi, Westal, Cheltenham. Jan. 1900 © private
      - Yellow [Winter] Jasmine, Westal, Cheltenham. New Year 1900 © private

| p10 | - Edward Wilson. c.1873 © CAGM |
| | - Soldiers. c.1877 © CAGM |

| p11 | - Edward Wilson. c.1882 © CAGM |
| | - Edward Wilson. c.1891 © CAGM |

| p12 | - Cheltenham from *The Crippetts*. c.1898 © CC |
| | - Coltsfoot. c.1898 © private |

| p13 | - Chaffinches. c.1900 © SPRI N:1870 |
| | - Crested Tit, Davos, Switzerland. March 1899 © private |

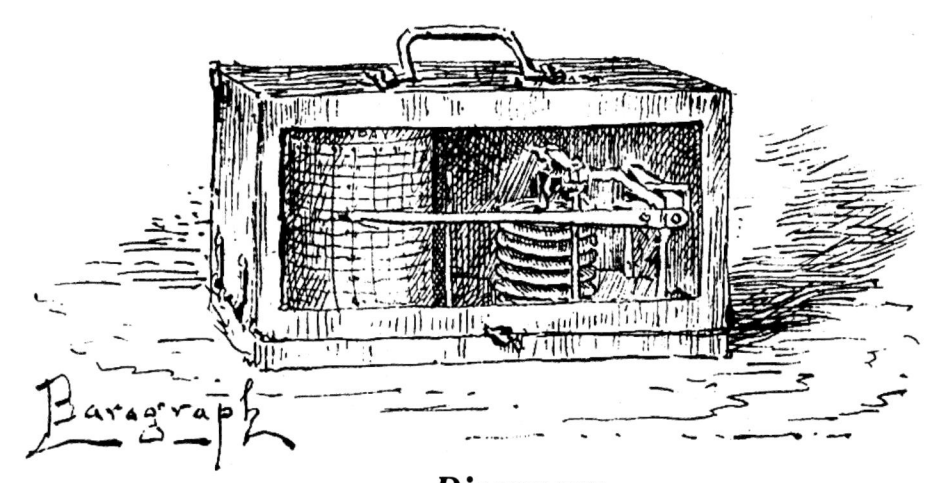

Discovery
The British National Antarctic Expedition: 1901-1904

p14 - *Discovery* with Parhelia. c.1905 © SPRI N:521

p15 - Adélie Penguin, running. Armitage 1905 © private
- Crest of the *Discovery* Expedition. c.1901 © private
- Sledging Notes, In Full Swing. Undated © SPRI N:1339
- Dolphin [Hourglass Dolphin] sketch. 01 Jan. 1902 © SPRI N:1406
- Sea Leopard [Leopard Seal] on Ice-floe. May 1902 © SPRI N:1425
- Blue Petrel. c.1902 © SPRI N:1676
- Emperor Penguin Foot. 14 Dec. 1903 © SPRI Y:56/27/6
- Iceberg. 25 Jan. 1902 © SPRI Y:61/4b

p16 - 'Billy', caricature of Edward Wilson by Lieut. Michael Barne. *South Polar Times*. Aug. 1902 © DHT
- 'He Preens', caricature of Edward Wilson for *South Polar Times* by Lieut. Michael Barne. Unpublished. c.1902 © private

p17 - Weddell Seals, Hut Point. Mar. 1903. Photograph by Edward Wilson [Skelton Index: Wilson 022]
    © SPRI p83/6/3/1/0154
- Edward Wilson field sketching during the *Discovery* Expedition. Undated © SPRI MS:750

p18 - Emperor Penguin with chick. Sept. 1903. Photograph by Edward Wilson [Skelton Index: Wilson 012]
    © SPRI P83/6/3/1/0144
- The Emperor Penguin rookery under the ice cliffs of Cape Crozier. Photograph by Edward Wilson [Skelton Index: Wilson 06]. Oct. 1903 © SPRI P:83/6/3/1 0138

p19 - Map of the European Antarctic Expeditions from a paper, *Antarctica*, given to the Cheltenham Natural Science Society by Edward Thomas Wilson. c.1901 © CAGM

p20 - Crab shrimp sp.. 05 Sept. 1901 © NMBL
- Pteropod. 10 Aug. 1901 © NMBL
- Shrimp sp.. 05 Sept. 1901 © NMBL
- Black-bellied Storm Petrel, field sketch. 01 Sept. 1901 © SPRI Y:2006/15/26
- Wilson's Storm Petrel, field sketch. 10 Sept. 1901 © SPRI Y:2006/15/30
- Sunrise, looking South. 09 Sept. 1901 © SPRI Y:2007/5/12

p21
- South Trinidad at dawn. Sept. 1901 © SPRI N:1322
- South Trinidad, Monument, Sugar Loaf and Noah's Ark. 13 Sept. 1901 © SPRI N:1323
- South Trinidad [Herald] Petrel sitting. 13 Sept 1901 © SPRI N:1649
- Head of South Trinidad [Herald] Petrel. Undated. Probably Sept. 1901 © SPRI N:1685

p22
- Cape Pigeon [Petrel] specimen. 21 Sept. 1901 © SPRI N:1697
- Cape Pigeon [Petrel] specimen. 21 Sept. 1901 © SPRI N:1698
- Cape Pigeon [Petrel] specimen. 21 Sept. 1901 © SPRI N:1699
- Cape Pigeon [Petrel]. Sketches at sea. 18-19 Sept. 1901 © SPRI Y:2006.15.28
- Darker and Larger Grey backed Collared Petrel. Sketches at sea. 18-19 Sept. 1901 © SPRI Y:2006.15.29

p23
- Fish caught on Agulhas Bank, South Africa. 15 Oct. 1901 © NHM
- South African Penguin sketch. *South Polar Times*. July 1902 © DHT
- Coast from Cape Town to Simon's Bay, White Sands, South Africa. 05 Oct. 1901 © SPRI Y:2007.5.13a
- Table Mountain, South Africa. 04 Oct. 1901 © SPRI Y:2009.14a

p24
- The Roaring Forties. 01 Nov. 1901 © private
- Sketch of Captain Scott. *Discovery*. 1901 © SPRI N:1333
- Sunset, Southern Ocean. Oct. 1901 © SPRI N:1945
- Sunset, South Indian Ocean. Oct. 1901 © SPRI N:1946
- Seabird sketches. 18 Oct. 1901 © SPRI Y:2006.15.32a

p25
- Pencil sketches of ice. 17 Nov. 1901 © SPRI N:1380/N:1381
- Antarctic Petrel specimen. Nov. 1901 © SPRI N:1659
- Antarctic Prion specimen Nos. 7 and 9. Head and feet. 16 Nov.1901 © SPRI N:1669

p26
- King Penguin heads, for *Discovery* Expedition Scientific Reports. 1907 © SPRI Y:67/5/13
- Crested [Royal] Penguin specimen 22. 22 Nov. 1901 © SPRI N:1501
- Antarctic Skua specimen 24. 22 Nov. 1901 © SPRI N:1740
- Schlegeli's [Royal] Penguin sketch. *South Polar Times*. July 1902 © DHT
- King Penguin sketch. *South Polar Times*. July 1902 © DHT
- King Penguin chicks sketch. *South Polar Times*. July 1902 © DHT

p27
- Wandering Albatross, sketches at sea. 1901 © SPRI N:1633
- Black-browed Albatross sketches at sea. 1901 © SPRI N:1639
- Light-mantled Sooty Albatross sketches at sea. 1901 © SPRI N:1645
- A present from New Zealand sketch. *South Polar Times*. June 1903 © DHT

p28
- Wandering Albatross. Undated. © SPRI N:418
- Black-browed [Campbell] Albatross, specimen 1. 28 Dec. 1901 © SPRI N:421
- Grey-headed Albatross, specimen 6. 29 Dec. 1901 © SPRI N:1644
- Light-mantled Sooty Albatross. Undated © SPRI N:1646/1
- Light-mantled Sooty Albatross. 'What a mouth'. Undated © SPRI N:1646/2

p29
- Iceberg. Undated © SPRI N:1329
- Second iceberg sighted after New Zealand. 04 Jan. 1901 © SPRI N:1349
- Porpoise [Hourglass Dolphin]. 01 Jan. 1901 © SPRI N:1406
- Immature Emperor Penguin, Pack Ice, specimen 28. 06 Jan. 1902 © SPRI N:1469
- Snow Petrel, Pack Ice, specimen 15. 07 Jan. 1902 © SPRI N:1652

p30
- Iceberg. Undated © AHAG 00406/64
- Sea Leopard [Leopard Seal] sketch. Undated © SPRI N:1424
- Crabeater Seal sketches. Undated © SPRI N:1428
- Ross's Seal sketches. Undated © SPRI N:1440
- Ross's Seal sketch. 06 Jan. 1902 © SPRI N:1442

p31
- Adélie Penguin sketches. Armitage 1905 © private
- The Two Sisters, Cape Adare. Armitage 1905 © private
- Adélie Penguin sketches, Cape Adare. Undated © SPRI N:1488
- Adélie Penguin feet, for *Discovery* Expedition Scientific Reports. 1907 © SPRI Y:67/5/14

p32
- The Two Sisters, Cape Adare. Undated © SPRI N:532
- The Two Sisters, Cape Adare. Undated © SPRI N:1285

p33	- Wilson's Storm Petrel, Cape Adare. 09 Jan. 1902 © SPRI N:1687
	- South Polar Skua chick, Cape Adare. 09 Jan 1902 © SPRI N:1737
	- South Polar Skua, specimen 78, Cape Adare. 09 Jan 1902 © SPRI N:1736
	- Antarctic Pack icescape. Undated © SPRI Y:52/60

p34	- Cape Wadworth, Coulman Island to Lady Newnes Bay. Undated © SPRI N:1387
	- Cape Anne, Coulman Island. 13 Jan. 1902 © SPRI Y:2006/24/1
	- Midnight Sun on Admiralty Range. 10 Jan. 1902 © SPRI Y:61/9

p35	- First view of Mt. Erebus. 19 Jan. 1902 © SPRI N:1803/84
	- Mt Erebus. 19 Jan. 1902 © SPRI Y:61/12
	- Pans of Pack Ice at midnight. 19 Jan. 1902 © SPRI Y:61/5

p36	- Field sketches, finished pencil sketches and water-colours of "A lane in the Great Ice Barrier". 28 Jan. 1902 © SPRI N:1803/54; N:1345; N:535; Y:61/11; Y:66/10

p37	- The Great Ice Barrier. Undated © AHAG 02458/83
	- Berg off the Barrier and Solar Rainbow. 29 Jan. 1902 © private
	- Tilted Bergs off the Barrier. 25 Jan. 1902 © RGS S00014199

p38	- King Edward VII Land. 31 Jan. 1902 © SPRI N:1298
	- Emperor Penguin, Farthest East. 31 Jan. 1902 © SPRI N:1464
	- King Edward VII Land. Undated [1905] © Her Majesty Queen Elizabeth II 452348
	- Disintegrating ice, East end of Barrier. Undated © SPRI N:1801/71

p39	- Berg recently tilted. 07 Feb. 1902 © SPRI N:1373
	- Stratified pieces. 28 Jan. 1902 © SPRI N:1382
	- Stratified pieces off King Edward's Land. 28 Jan. 1902 © SPRI N:1777
	- Pieces of Barrier Ice. 28 Jan. 1902 © SPRI N:1803/47
	- Flying balloon. 03 Feb. 1902. *South Polar Times*. June 1902 © DHT
	- Western End of Ice Barrier with Mt. Terror. Undated © SPRI Y:79/15/2

p40	- *Discovery* in Winterquarters, looking West. Undated © SPRI N:1803/79
	- Hut at Hut Point, with dogs and camp. Undated © NMBL
	- Hut at Hut Point, *Discovery's* Winterquarters 1902-1904. Undated © SPRI Y:2007/5/14

p41	- Shackleton, Wilson and Ferrar setting out for White Island with Pram. 19 Feb. 1902 © BAL 375986
	- Shackleton and Ferrar setting camp, White Island sledging trip. Undated © BAL 375987
	- View from Crater I on Island I [White Island], looking South and South West. 20 Feb. 1902 © SPRI MS:366/12/5

p42	- Beaked Whales in Terror Bay. 25 Feb. 1902 © NHM Z88ffw
	- Sunset looking across McMurdo Sound. Autumn 1902 © SPRI Y:2007/5/21

p43	- View from Hut Point Peninsula to Winterquarters and the Royal Society Range. 28 Mar. 1902 6pm © SPRI N:1803/105
	- View from Hut Point Peninsula to Winterquarters and the Royal Society Range. 28 Mar. 1902 © SPRI N:1278

p44	- View of Mt. Lister and Royal Society Range, new ice forming. 17 Apr. 1902 © SPRI N:1293
	- View of Mt. Lister and Royal Society Range, sunset. 11 Apr. 1902 4.30pm © SPRI N:522
	- View of Mt. Lister and Royal Society Range, sunset. Undated © RGS S0021470
	- View of Mt. Lister and Royal Society Range, sunset. Undated © RGS S0017162

p45	- Sledging in Apr., Camping after Dark. Depot laying Journey. Apr. 1902 © SPRI N:1394
	- Penguin Hunt at Winterquarters. Apr. 1902 © SPRI N:1344
	- Emperor Penguins. 17 Apr. 1902 © SPRI N:1462
	- McCormick's [South Polar] Skuas. Apr. 1902 © SPRI N:1749
	- Sketch of *Discovery* in Winterquarters. Undated © SPRI N:1803/78

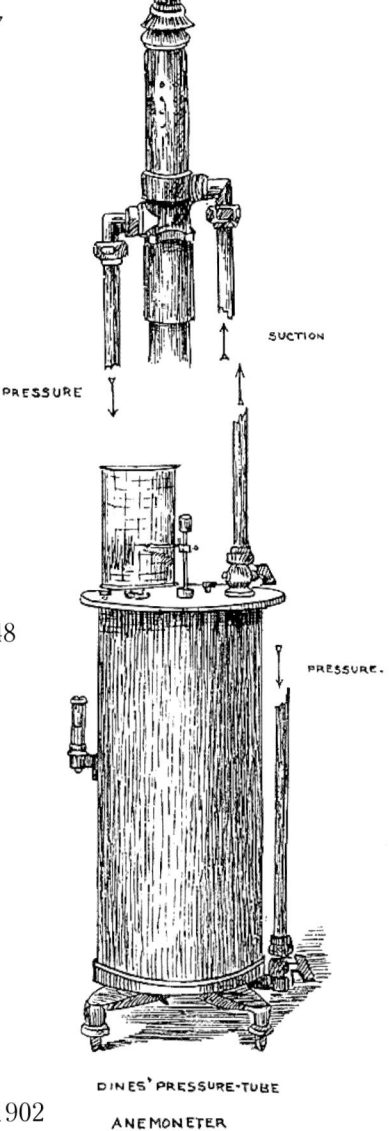

DINES' PRESSURE-TUBE ANEMOMETER

p46
- *Discovery* in Winterquarters. Aug. 1902. © SPRI N:1803/77
- Frontispiece. *Discovery* in Winterquarters. *South Polar Times*. Aug. 1902 © DHT
- *Discovery* in Winterquarters. 1903. Dated 1903 © SPRI N:542
- *Discovery* in Winterquarters. 1903. Dated 1903 © SPRI Y:2007/5/15

p47
- On the way to the screen in a blizzard. Undated © Armitage 1905 p216. private
- Frontispiece. Trip to the meteorological observatory in a blizzard. *South Polar Times*. May 1902 © DHT
- Taking observations in a blizzard. Undated © CAGM 1930.64

P48
- Auroral Arc and Curtain above glare in East. 05 July 1902 0.30am © SPRI N:501
- Double Auroral Arc Vertical rays in Upper Arc. 29 Aug. 1902 2am © SPRI Y:2006.25.2
- Auroral Curtains, looking North from Winterquarters. 05 July 1902 1am © SPRI N:1803.151

p49
- Auroral Streamers. 09 Apr. 1902 2.25am © SPRI Y:2006.25.1
- *Discovery* in Winterquarters with Aurora. Undated © SPRI N:543
- Auroral Corona, looking to the zenith, McMurdo Sound. 08 Apr. 1903 2am © SPRI N:1803/142

p50
- The Gap in winter moonlight. Undated © SPRI N:541

p51
- Shackleton's sledging flag. *South Polar Times*. 1902 © SPRI N:1795
- Wilson's sledging flag. *South Polar Times*. 1902 © SPRI N:1796
- Scott's sledging flag. *South Polar Times*. 1902 © SPRI N:1797
- Silhouette of Captain Scott. *South Polar Times*. June 1902 © DHT
- Title Page. *South Polar Times*. Apr. 1902 © DHT
- Frontispiece. Cape Crozier sledging party. *South Polar Times*. Apr. 1902 © DHT
- April dog sledging on the Barrier. *South Polar Times*. Apr. 1902 © DHT
- Mid-winter dinner menu. *South Polar Times*. June 1902 © DHT

p52
- Fire-glow on the smoke of Mt. Erebus. 1902 © SPRI Y:54/25/1

p53
- Looking North in McMurdo Strait. 26 July 1902 mid-day © RGS S0010530
- Royal Society Range with Mt. Lister - Western Mountains, McMurdo Strait. 14 Aug. 1902 mid-day © SPRI N:1286

p54
- Eschenhagen Magnetograph. *South Polar Times*. May 1902 © DHT
- Magnetograph instruments. *South Polar Times*. May 1902 © DHT
- Isopods. *South Polar Times*. June 1902 © DHT
- Worms. *South Polar Times*. June 1902 © DHT
- *Notothenia* sp.. *South Polar Times*. June 1902 © DHT
- Red snow algae cells, sporing. *South Polar Times*. Aug.1902 © DHT
- Meteorological Screen and Magnetic Hut. *South Polar Times*. Apr. 1902 © DHT
- *Tubularia* sp., D net, Hut Point. Sept. 24 1902 © Cheffins
- *Periphylla periphylla*, Helmet Jellyfish, McMurdo Strait.01 Aug. 1902 © NHM 030907

p55
- Erebus, Terror and Rocky headland from Tortoise Rock. Undated © SPRI N:464
- Erebus, Terror and Rocky headland from Tortoise Rock. Undated © private
- Setting a fish trap, Mt Erebus and Mt. Terror from the Turtle Back Island. Undated © private
- The Biologist at work in his shelter. Armitage 1905 © private

p56
- Looking towards Crater Hill. Early Aug. 1902 mid-day © SPRI N:1269
- McMurdo Sound, sun's approach. Aug. 1902 © SPRI N:1276

p57
- Emperor Penguin chick, taken alive, the largest. Oct. 1902 © SPRI N:1470
- Emperor Penguin chick, taken alive, the middle sized one. Oct. 1902 © SPRI N:1473
- Emperor Penguin chick, found dead, the smallest. Oct. 1902 © SPRI N:1477

"SLEDGING HARNESS"—FOR MEN.

p58	- Sledging sketches. Undated © SPRI N:1340
- Sledging sketches. Undated © SPRI N:1341
- 3 in a sleeping bag sketch. Undated © SPRI N:1342
- Sledging with sail. Undated © SPRI N:1400
- Antarctic sledging. 1903 © SPRI N:507
p59	- Taking in the cooker. Undated © private
- Pitching a tent in wind sketch. Undated © SPRI N:1343
- Pitching a tent in wind. Undated © private
p60	- 3 men in a pyramid tent. Undated © SPRI N:508
- 3 men in a pyramid tent. Undated © SPRI N:1395
- 3 men in a pyramid tent. Undated © SPRI N:1396
p61	- Portrait of a sledge dog, Joe. *South Polar Times*. June 1903 © DHT
- Portrait of a sledge dog, Bismarck. *South Polar Times*. June 1903 © DHT
- Portrait of a sledge dog, Jim. *South Polar Times*. June 1903 © DHT
- Portrait of a sledge dog, unidentified. *South Polar Times*. June 1903 © DHT
- Dog Camp on the Great Ice Barrier with Parhelia and Perihelia. Undated © SPRI N:526
- Preparatory sketch to calculate correct angles of Parhelia. Undated © SPRI Y:54/25/3
p62	- Mt. Longstaff, 13,000 feet, Farthest South of all. 1902 © SPRI N:473
- Farthest South land. 1902. © SPRI N:1294
- Mt. Longstaff, Farthest South. Undated [1905] © Her Majesty Queen Elizabeth II 452347
p63	- Shackleton, Wilson and Scott return North past Shackleton Inlet and Cape Wilson, with Mt. Markham in the background. Undated © AHAG
- Farthest South Camp with Parhelia. *South Polar Times*. June 1903 © DHT
- Sketch of Mt. Markham, Shackleton Inlet and Cape Wilson. Farthest South. Dec. 1902 © SPRI N:1303
p64	- The Last of the Dogs. Undated © private
- 3 men skiing. Undated © CAGM
- Fog-bow on the Great Ice Barrier. Undated © SPRI N:506
p65	- Sketches of Weddell Seals, McMurdo Strait. Mar. 1903 © SPRI N:1407; N:1408; N:1410; N:1413; N:1415; N:1416; N:1417; N:1421
p66	- *Discovery* in Winterquarters, looking North. Undated © SPRI N:1240
- *Discovery* in Winterquarters, looking North. 28 Apr. 1903 noon © SPRI N:1242
p67	- Castle Rock, near Winterquarters. Undated © RGS S0022126
- Castle Rock, from the ski slopes. 03 Aug. 1903 2pm © SPRI N:1271
- Crater Hill from Harbour Heights looking South East. 08 May 1903 noon © SPRI N:1279
p68	- Earth Shadows. Undated © SPRI N: 499
- Earth Shadows. 24 Apr. 1903 © SPRI N:1288
- Earth Shadows. Undated © SPRI Y:54/25/2
p69	- Mt. Erebus from the ski slope. 26 Apr. 1903 © SPRI N:518
- Mt. Erebus and Castle Rock from Harbour Heights. 16 Aug. 1902 noon © SPRI Y:2006/24/2
- Mt. Erebus from Crater Hill, showing unusual form of smoke. 26 Apr. 1903 © SPRI N:1801/68
p70	- The Aurora Australis. Undated © SPRI 1386
p71	- 'A Merry Mid-winter'. *South Polar Times*. June 1903 © DHT
- Polar Hieroglyphs. *South Polar Times*. June 1903 © DHT
- Summer Fishing. *South Polar Times*. Apr. 1903 © DHT
- Sledging Gear/Antarctic Cavalry. *South Polar Times*. Apr. 1903 © DHT
- Frontispiece. Ship in Ice. *South Polar Times*. Apr. 1903 © DHT
- The deck of *Discovery* in winter. *South Polar Times*. Aug. 1903 © SPRI

p72	- *Discovery* in Winterquarters, looking North. 23 July 1903 noon © SPRI 1243
	- *Discovery* in Winterquarters, looking North. Undated © private
	- Looking North East about noon, Mt. Terror, Castle Rock and Danger Slopes. 20 July 1903 © SPRI 1277

p73
- Chiton, from #10 hole, 136 fathoms. 18 June 1903 © NMBL
- Ice Crystals from deck houses on ship. Undated © SPRI N:1801/58
- Ice Crystal. Undated © SPRI 1803/152
- Fish taken from stomach of Weddell Seal. 27 Sept. 1903 © NHM Z88fw
- Mt. Erebus smoke showing upper current of air. Undated © SPRI N:1297
- Topographical drawing. View from Harbour Heights looking South West. Undated © SPRI N:544

p74
- *Discovery* in Winterquarters. 13 Sept. 1903 © SPRI N:1241
- South Polar (McCormick's) Skuas near nest. Undated © SPRI N:494

p75
- Opalescent Alto Stratus and snowdrift *Discovery* in Winterquarters, McMurdo Sound, looking North. 17 Aug. 1903 1-2pm © SPRI N:1245
- Looking West to Mt. Lister, McMurdo Strait. 16 Aug. 1903 2pm © SPRI N:1275
- Looking South, Mt. Discovery. 03 Aug. 1903 1pm © SPRI N:1281

p76
- The Emperor Penguin Rookery at Cape Crozier. 1903 © SPRI N:491

p77
- Emperor Penguin chick sleeping. Sept. 1903 © private
- Sketch of Emperor Penguin and chick. Undated © SPRI N:1456
- Emperors fighting over live chick. 13 Sept. 1903 © SPRI N:1457
- Emperor approaching dead chick. Undated © SPRI N:1457
- Emperor Penguin showing method of holding young chick. 13 Sept. 1903 © SPRI N:439
- Emperor Penguin feeding a chick. 13 Sept. 1903 © SPRI N:444
- Emperor Penguin with dead chick. 13 Sept. 1903 © SPRI N:446

p78
- Emperor Penguins and chick. Oct. 1903 © RGS S0010531
- The Emperor Penguin Rookery at Cape Crozier. 1903 © BAL 375984
- Emperor Penguins. 1904 © CAGM 2000.30

P79
- Emperor Penguin Rookery at Cape Crozier. 1903 © SPRI N:1460
- Emperor Penguin chick, No 13, preserved in spirit. 13 Sept. 1903 © SPRI N:441
- Emperor Penguin chick, No 5, preserved in formalin. 13 Sept. 1903 © SPRI N:1471

p80
- Emperor Penguin chick. Undated © SPRI N:447
- Emperor Penguin chick. Undated © SPRI N:1449
- Emperor Penguin chick running from parents. 26 Oct. 1903 © SPRI N:1450
- Emperor Penguin chick, from life. Nov. 1903 © SPRI N:1455
- Emperor Penguin chick. Undated © SPRI N:1459
- Emperor Penguin eggs, draft plate for Scientific Reports. 1904 © SPRI N:1480

p81
- Glacier Tongue coming to within 2 miles of coast. 04 Dec. 1903 © SPRI N:1801/8
- Granite Cliffs, West side of McMurdo Strait. 09 Dec. 1903 9am © SPRI N:1801/21
- Western Range heights, from last outward camp. 06 Dec. 1903 © SPRI N:1801/15
- Glacial Gravel Piers, Western Journey. Undated © SPRI N:1801/14

p82 - The Ecstatic Position assumed by *Pygoscelis adeliae*. Undated © SPRI N:1485
- Field sketches of the Adélie Penguin, Cape Royds. Jan. 1904 © SPRI N:1486 and N:1487
- The Coming of the Relief Ships. 1904 © SPRI Y:69/10/2/35
- *Discovery* in Winterquarters Bay, with Vince's Cross. 1904 © CAGM 1964.181

p83 - Vince's Cross at Hut Point. Feb. 1904 © SPRI N:1289
- *Terra Nova* and *Morning* departing McMurdo Sound. 19 Feb. 1904 midnight © SPRI N:1803/81
- *Terra Nova* passing the Possession Islands. © SPRI N:538

p84 - Mt. Minto and Mt. Adam, heights of Admiralty Range, seen through the gap in the Cape Adare Cliffs. 24 Feb. 1904 © RGS S0021473
- Small tabular berg. Mar. 1904 © SPRI N:1330
- Southern Cross Hut, Cape Adare. 25 Feb. 1904 © SPRI N:1282
- *Thalassoica antarctica*, Antarctic Petrel, Balleny Islands. 02 Mar. 1904 © SPRI N:1661

p85 - New Zealand Falcon, Port Ross, male. 23 Mar. 1904 © SPRI N:1935
- New Zealand Falcon, Port Ross, female. 16 Mar. 1904 © private (original now at Canterbury Museum, New Zealand)
- *Sterna frontalis* [White-fronted Tern], Port Ross, skin 120. 25 Mar. 1904 © SPRI N:1722
- Silvereye, Port Ross. 20 Mar. 1904 © SPRI N:1937
- *Platycercus auriceps*, Kakariki, Port Ross. 26 Mar. 1904 © SPRI N:1941

p86 - Maori Whare, Tokaanu, New Zealand. 08 May 1904 © SPRI Y:2006/15/41c
- Pukeko, Wanganui, New Zealand. Undated [1904] © private
- Unlabelled. [Magellan Strait]. Undated [1904] © SPRI N:537
- Female Kelp Goose. Undated [1904] © private
- Steamer Duck, Falkland Islands. Undated [1904] © private
- St. Michael's Church, Azores. 02 Sept. 1904 © CC

p87 - Weddell Seal, for *Discovery* Expedition Scientific Reports. 1907 © SPRI Y:79/19
- Immature Emperor Penguin Studies, for *Discovery* Expedition Scientific Reports. 1907 © SPRI Y:67/5/5
- McCormick [South Polar] Skua Studies, for *Discovery* Expedition Scientific Reports. 1907 © SPRI Y:67/5/16

The Natural History of the British Isles: 1904-1910

p88 - Red Grouse. Undated © private

p89 - Immature Dormouse. Poynetts. 11 Sept. 1905 © NHM T36133
- Portrait of a pony. c.1905 © private (original now at Bushey Museum)
- Larch cones. *The Crippetts*. 1909 © private
- Red Kite, preparatory plate for Yarrell's *British Birds*. 1905-1910 © private
- Kingfisher, preparatory plate for Yarrell's *British Birds*. 1905-1910 © SPRI N:1851
- Hare sketches for Barrett Hamilton's *British Mammals*. 1905-1910 © NHM T36154

p90 - Mr and Mrs Reginald Smith, with Oriana Wilson and Captain Scott, picnicking near Cortachy. c.1907. Photograph by Edward Wilson © Wilson Collection, CAGM
- Edward Wilson. Photograph by Elliot and Fry. c.1910 © Wilson Collection, CAGM

p91 - Two Adélie Penguins, before and after. "Drawn by Ted on the plates mother provided for our lunch on the journey to Kerry". 1905 © Wilson Collection, CAGM

p92 - Moonlight on a Frozen Sea. Plate for Scott's *The Voyage of the Discovery*. 1905 © SPRI Y:2006/15/42
- Poster for the Bruton Gallery Exhibition. 1904 © Wilson Collection, CAGM

p93 - Valentia Island from Anascaul, Ireland. Aug. 1905 © private
- Poster for a talk by Edward Wilson. Feb.1910 © Wilson Collection, CAGM

Terra Nova
The British Antarctic Expedition: 1910-1913

p94   - Lunar Corona, Cape Evans, McMurdo Sound. 1911 © RGS S0021476

p95   - Taking a Sounding. Undated © private
- *Terra Nova*. Armitage 1905 p282 © private
- Expedition Crest. Undated © private
- Head of a Black-browed Albatross. Dec. 1910 © SPRI N:422
- Mt. Lister. 16 Sept. 1911 evening © SPRI N:435
- Portrait of Captain Scott. Undated © SPRI N:1392
- Igloo at Cape Crozier. *South Polar Times*. Sept. 1911 © SPRI

p96   - Edward Wilson bird-watching aboard *Terra Nova*. 1910 © Wilson Collection, CAGM
- Wilson and Pennell salting seal skins aboard *Terra Nova*. 27 Dec. 1910 © SPRI p2005/5/0049

p97   - Edward Wilson after sledging. Undated © SPRI p2005/5/1423
- Bowers, Wilson and Cherry-Garrard, departing for the Winter Journey to Cape Crozier. June 1911 © SPRI p2005/5/0452

p98   - 'Our Bill', caricature of Edward Wilson by Lillie. *South Polar Times*. Sept. 1911 © SPRI Y:54/24/2
- Edward Wilson with his pony, Nobby. Oct. 1911 © SPRI p2005/5/1646

p99   - The Norwegian Tent at the South Pole, Wilson sketching. Photograph by Bowers. Jan. 1912 © SPRI p2005/5/1441
- Wilson, Scott, Evans, Oates and Bowers at the South Pole. 18 Jan. 1912 - the last known photograph taken by Edward Wilson. © SPRI p2005/5/1475

p100   - *Procellaria pelagica*, [*Oceanodroma castro*] Madeiran Storm Petrel. 09 July 1910 © SPRI N:1647
- *Bulweria bulweri* [Bulwer's Petrel], flight pattern. 06 July 1910 © SPRI N:1678
- Sunset No 1. Clouds beginning to form. 01 July 1910 © SPRI Y:67/1/4
- Rain squall. Looking South. 12 July 1910 8am. © SPRI Y:67/1/7
- Sunset. West. 13 July 1910 about 6pm © SPRI Y:67/1/14

p101   - South Trinidad Island, one of the tops. Undated © SPRI N:1320
- South Trinidad Petrel, young male, shot. 28 July 1910 © SPRI N:1684
- Fairy Tern, shot off shore. 26 July 1910 © SPRI N:1720
- Frigate bird, shot off shore. 26 July 1910 © SPRI N:1731
- Frigate bird feet, No's 4 and 5. Undated © SPRI N:1733

p102   - Diving Petrel sp.. 13 Aug. 1910 © SPRI N:1648
- Probable Atlantic Petrel. 22 July 1910 © SPRI N:1677
- Soft-plumaged Petrel. 03 Aug. 1910 © SPRI N:1683
- Œstrelata mollis, [*Pterodroma mollis*] Soft-plumaged Petrel. 04 Aug. 1910 © SPRI Y:2006/15/45
- Young Wandering Albatross. 12 Aug. 1910 © SPRI Y:2006/15/46
- Young South African Penguin, Cape Colony. Undated © SPRI Y:2006/15/55
- False Bay, South Africa. 1910 © SPRI Y:67/1/12

p103   - Preparatory plate for Southern Ocean Seabirds, *Diomedea exulans*, Wandering Albatross. Most sketches are made from RMS *Corinthic*. Sept. 1910 © SPRI N:1634

p104   - Bulwer's Petrel seen from RMS *Corinthic*. 25 Sept. 1910 © SPRI N:1679
- Bird notes from RMS *Corinthic*. 13 Sept. 1910 © SPRI MS:234/4
- Sunset. South Indian Ocean, South Lat. 42°. 12 Sept. 1910 © SPRI Y:67/1/10
- Pencil sketch of Frogmouth, Melbourne. Oct. 1910 © SPRI MS:234/4
- Pencil sketch of Blue Wren, Melbourne. Oct. 1910 © SPRI MS:234/4

p105   - Dolphin sp.. 18 Oct. 1910 © NHM 058526
- Buller's Albatross. Oct.1910 © SPRI N:1628
- Giant Petrel sketches. 26 Oct. 1910 © SPRI Y:2006/15/47 and 48
- Wandering Albatross settling. 26 Oct. 1910 © SPRI Y:2006/15/49
- Gum Tree, Middleton, New Zealand. Oct. 1910 © SPRI Y:2006/15/51

| p106 | - Tabular Iceberg to East. 09 Dec. 1910 © SPRI N:1953
- A Berg in the Pack, Ross Sea. Dec. 1910 © CAGM 1930:63
- Preparatory sketch in the Pack Ice. 21 Dec. 1910 l0pm © SPRI N:1954
- Overturned Berg. 09 Dec. 1910 © SPRI N:1952 |
|---|---|
| p107 | - In the Pack Ice. 21 Dec. 1910 l0pm © SPRI N:412
- Pack Ice and Bergs in Ross Sea. Dec. 1910 © private |
| p108 | - Adélie Penguins in the Pack Ice. Undated © AHAG 407/64 |
| p109 | - Berg in the Pack Ice with Emperor Penguins. 20 Dec. 1910 10am © SPRI N:460
- Antarctic Petrels resting on an Iceberg. 19 Dec. 1910 © SPRI N:461 |
| p110 | - *Pagodroma nivea*, Snow Petrels, Ross Sea. 1910 © SPRI N:1932 |
| p111 | - Iceberg in the Pack Ice. 21 Dec. 1910 10.10am © SPRI N:413
- Tabular Berg. 09 Dec. 1910 5am © SPRI N:428 |
| p112 | - Whale bird, Prion sp.. Undated © SPRI N:1671
- Light-mantled Sooty Albatross. Undated © SPRI Y:2006/15/63
- Crabeater Seal sketches. 18 Dec. 1910 © NHM 058528
- Adélie Penguin sketches. Dec. 1910 © SPRI MS:234/4 |
| p113 | - Light-mantled Sooty Albatross. Undated © SPRI Y:2006/15/64
- Crabeater Seal sketches. 27 Dec. 1910 © NHM 058529
- Emperor Penguin sketches. 19 Dec. 1910 © SPRI MS:234/4
- Adélie Penguin sketches. Dec. 1910 © SPRI MS:234/4 |
| p114 | - Pack Ice, Ross Sea. Jan. 1911 © SPRI N:469 |
| p115 | - Pack Ice, Ross Sea. Jan. 1911 midnight © private
- Beaufort Island, Ross Sea. 04 Jan 1911 1am © SPRI N:416 |
| p116 | - The Great Ice Barrier looking East from Cape Crozier. 04 Jan. 1911 © SPRI N:493
- The Barrier at Cape Crozier 04 Jan. 1911 © SPRI N:550 |
| p117 | - The Barrier looking East from off Cape Crozier. 04 Jan. 1911 © SPRI N:1398
- Junction of Barrier and Cape Crozier. 04 Jan. 1911 © SPRI N:1399 |
| p118 | - Cave in the Ice Barrier, Cape Crozier. 04 Jan. 1911 © SPRI Y:2007/5/17 |
| p119 | - Ponting and the Whales, cartoon. *South Polar Times*. Oct. 1911 © private
- Captain Oates with a pony, cartoon. *South Polar Times*. Oct. 1911 © private
- Cape Evans with the Ramp and Mt. Erebus. *South Polar Times*. June 1911 © private
- Sledge hauling on ski. Mar. 1911 © private |
| p120 | - Paraselena. Cape Evans, McMurdo Sound. 15 Jan. 1911 9.30pm © SPRI 478 |
| p121 | - In the teeth of a blizzard. *South Polar Times*. Oct. 1911 © SPRI
- The Rescue from the Sea ice. 01 March 1911. *South Polar Times*. Sept. 1911 © SPRI
- Down a Crevasse. Undated © private |
| p122 | - Hut Point. 1911 © CAGM 1968.190
- Hut Point, McMurdo Sound. 07 Apr. 1911 © SPRI N:423
- Hut Point, McMurdo Sound. 07 Apr. 1911 © SPRI N:1802/9 |
| p123 | - New ice at Hut Point. 29 Mar. 1911 noon © SPRI N:434
- Hut Point, frost smoke and new ice. 03 Apr. 1911 6pm © SPRI Y:2006/4/1
- Hut Point. 29 Mar. 1911 noon © SPRI N:525
- Hut Point. 03Apr. 1911 6pm © SPRI N:1802/7 |
| p124 | - Hut Point from the top of Observation Hill. 01Apr. 1911 5.30pm © SPRI N:509 |

p125	- Hut Point from the top of Observation Hill. 01 Apr. 1911 5.30pm © SPRI N:496 - Field sketch from the top of Observation Hill. 01 Apr. 1911 5.30pm © SPRI N:510 - Field sketch from the top of Observation Hill. 13 Apr. 1911 © SPRI N:1802/11
p126	- Mt. Erebus and Castle Rock from Observation Hill. 31 Mar. 1911 © BAL 421598 - Mt. Erebus and Castle Rock from Observation Hill. 31 Mar. 1911 © SPRI MS:750/171
p127	- Mt. Erebus. 02 Apr. 1911 6pm © SPRI N:453 - Mt. Erebus. 28 Apr. 1911 © SPRI N:475
p128	- Mt. Erebus from Hut Point. Mar. 1911 © SPRI N:470 - Mt. Erebus from Hut Point. Mar. 1911 © RGS S0022127
p129	- Hut Point Ski Slope. 30 Mar. 1911 sunset 7pm © BAL 375981 - Looking across McMurdo Sound from Hut Point. 29 Mar. 1911 noon © SPRI N:487
p130	- Mt. Discovery, with open leads in new ice. 26 Mar. 1911 7pm © SPRI N:426 - Mt. Discovery, sunset over new ice. 25 Mar. 1911 © CAGM 1930:61 - Pencil sketches and notes for sunsets over Mt. Discovery. Mar. 1911© SPRI N:1802/3 and 4
p131	- Observation Hill from Hut Point. 09 Apr. 1911 6pm © CAGM 1930:65 - Observation Hill from Fodder Depot. 14 Apr. 1911 2pm © SPRI N:486 - McMurdo Sound. 28 Mar. 1911 10pm © SPRI N:490
p132	- Frost Smoke, Looking North, Easter Sunday. 16 Apr. 1911 © SPRI N:457 - Sunset, Looking West, from Hut Point. 01 Apr. 1911 © SPRI N:514
p133	- Sunset. 31 Mar. 1911 © SPRI N:425 - Sunset. 10 Apr. 1911 6pm © SPRI 438 - Sketch of sunset looking West from Hut Point. 01 Apr. 1911 © SPRI N:1802/8
p134	- Looking North from Hut Point, very young ice and frost smoke. 02 Apr. 1911 7pm © private - Frost Smoke and new ice in McMurdo Sound. 30 Mar. 1911 7pm © SPRI N:429
p135	- Sunset, McMurdo Sound. 13 Apr. 1911 © AHAG 02850/87 - Down the Hutton Cliffs. 21 Apr. 1911. *South Polar Times*. June 1911 © private - Noon, Looking South. 29 Apr. 1911 © SPRI N:513
p136	- Berg off Cape Evans, Last day of the Sun. 23 Apr. 1911 © SPRI N:1401 - Cape Barne from Cape Evans. 28 Apr. 1911 noon © private
p137	- Barne Glacier from the Ramp. 30 Apr. 1911 1pm © SPRI N:414 - Frost smoke at Sunset, Easter Sunday. 16 Apr. 1911 © SPRI N:449
p138	- Paraselena, Cape Evans. 13 May 1911 8am © SPRI N:466 - Paraselena, McMurdo Sound. 20 June 1911 10.30am © SPRI N:471
p139	- Parasites in gut of *Trematomus* sp., Cape Evans. 06 May 1911 © NHM 057204 - *Trematomus* sp., Cape Evans. Fish Drawing No.1. 06 May 1911 © CC - *Trematomus* sp., Cape Evans. Fish Drawing No.2. 06 May 1911 © CC - *Trematomus* sp., Cape Evans. Fish Drawing No.3. 08 June 1911 © CC
p140	- Birdie Bowers, reading the thermometer on the Ramp. 06 June 1911 © BAL CH375983

p141   - Ice Crystals from glacier cave in South Bay. 28 Apr. 1911 © SPRI N:1801/61
            - Mt. Erebus at noon. 28 May 1911 © SPRI N:436
            - Cape Evans Hut. 15 May 1911 © SPRI N:1802/23
            - Birdie at 'Bertram', reading the thermometer on the Ramp. June 1911 © SPRI N:1802/28

p142   - Mid-Winter Dinner Menu. 22 June 1911. *South Polar Times*. Sept. 1911 © SPRI

p143   - Sledging flags, Oates and Bowers. *South Polar Times*. June 1911 © SPRI
            - Captain Scott, silhouette. *South Polar Times*. June 1911 © SPRI
            - Hieroglyph from Antarctic Archives. *South Polar Times*. June 1911 © SPRI
            - Frontispiece. *South Polar Times*. Sept. 1911 © SPRI
            - Bunk cartoons. *South Polar Times*. Sept. 1911 © SPRI
            - Sledging flag, Cherry-Garrard. *South Polar Times*. Oct. 1911 © SPRI

p144   - Erebus Slopes and the Ramp from Cape Evans. June 1911 6pm © SPRI N:407

p145   - 3 men sledge-hauling in a blizzard. Undated © AHAG 00410/64
            - Diagram of Oriana Hut and camp at Cape Crozier. Undated © SPRI MS:505/1
            - This is the hut that Cherry built. *South Polar Times*. Sept. 1911 © SPRI
            - 3 men stumbling in a blizzard. Undated © SPRI Y:2007/5/24

p146   - Cape Evans, Looking North. 08 Aug. 1911 3pm © SPRI N:527
            - Looking West from Cape Evans. Aug. 1911 © SPRI N:472

p147   - Cape Royds and Cape Barne from Cape Evans, part of the Archer Berg. 17 Aug. 1911 © SPRI Y:2007/5/23
            - Emperor Penguin Gut, cysts and parasites. 16 Aug. 1911 © NHM 057202
            - Weddell Seal Gut, with parasites. 09 Aug. 1911 © NHM 057203

p148   - Iridescent clouds, looking North from Cape Evans. 09 Aug. 1911 © SPRI N:433
            - Iridescent clouds, looking North from the Ramp on Cape Evans. 09 Aug. 1911 © SPRI N:488

p149   - Mt. Erebus. 13 Aug. 1911 10am © SPRI N:480
            - Spiral Cirrostratus Cloud. 11 Aug. 1911 © SPRI N:1388

p150   - Mt. Discovery and Inaccessible Island. 14 Aug. 1911 1pm © SPRI N:485
            - Inaccessible Island and Bergs, looking South from Barne Glacier. 13 Aug. 1911 © AHAG 00408/64

p151   - Mt. Erebus from Cape Evans. 12 Aug. 1911 10am © SPRI N:476
            - Mt. Erebus. 08 Sept. 1911 4pm © SPRI N:505

p152   - Shadow of Mt. Erebus and its smoke cloud. 09 Sept. 1911 9am © SPRI N:465
            - Mt. Erebus from Cape Evans. 13 Sept. 1911 © SPRI N:1390

p153   - An Iceberg off Cape Evans. 01 Sept. 1911 4.30pm © CAGM 1962:82
            - Cave berg off Cape Evans. 01 Sept. 1911 5.30pm © SPRI 528

p154	- Parhelion from the Ramp. 14 Sept. 1911 © RGS S0017161
- A Prismatic Parhelion. 11 Sept 1902 © RGS S0021472
p155	- Mt. Erebus. Undated © AHAG 00409/64
- Mt. Erebus. 19 Sept. 1911 © SPRI N:492
p156	- Mt. Discovery and Inaccessible Island from the Ramp. 14 Sept. 1911 5.30pm © SPRI N:483
- Royal Society Range. 14 Sept. 1911 5.30pm © SPRI N:482
p157	- Brown Island. Sept. 1911 © SPRI N:450
- Iceberg stranded off Cape Evans. 13 Sept. 1911 © SPRI Y:2004/4/2
- Royal Society Range. 14 Oct. 1911 noon © SPRI N:454
p158	- Field sketches from the Great Ice Barrier. Nov. 1911 © SPRI MS:797/1/BJ
p159	- Field sketches from the Beardmore Glacier. Dec. 1911 © SPRI MS:797/1/BJ
p160	- Field sketches from the South Pole. Jan. 1912 © SPRI MS:797/1/BJ; N:546; N:1988
p161	- Field sketches from the Polar Plateau and Beardmore Glacier. Feb. 1912 © SPRI MS:797/1/BJ

In Memoriam - One Hundred Years On: 1912-2012

p162	- Frigate bird specimens from South Trinidad at the British Natural History Museum. Type specimen for *Fregata ariel wilsoni*. 2011 © private/NHM
- The Christening Mug of Edward Wilson. 2011 © private
p163	- Specimen of *Glossopteris* collected by Edward Wilson on the Beardmore Glacier and carried on the sledge until he died. 1912 © NHM
- Scout Patrol Badge for 'Wilson Patrol'. c.1960-67 © private
p164	- Statue of Edward Wilson, Cheltenham. 2011 © Adrian Pingstone, Wikimedia Commons
- Edward Wilson memorial window, Cheltenham College. 2011 © CC
- Edward Wilson memorial window, Copthorne School. 2011 © Copthorne School
- The unveiling of the statue to Edward Wilson, Cheltenham. 1914 © Wilson Collection, CAGM
p165	- The Emperor Penguin eggs collected on the Winter Journey. 2011 © private/NHM
- The Wilson Room, Cheltenham Museum. c.2000 © CAGM
- South Pole sketches in the Royal Collection. 2011 © Her Majesty Queen Elizabeth II
- A balaclava worn by Edward Wilson. 2011 © private
- The new Polar Museum, SPRI, Cambridge, showing Wilson artefacts. 2011 © private/SPRI
p166	- Images of book covers relating to Edward Wilson. 2011 © private
- Prospectus from Edward Wilson Primary School. 2011 © Edward Wilson Primary School
- Photograph of Harold Warrender as Edward Wilson in the famous film, *Scott of the Antarctic*. 1948 © private
p167	- Used postage stamps, cigarette cards, postcards, pins, tins. Undated © private

End-pages

p168	- Emperor Penguins. Undated © SPRI N:443
- Adélie Penguin sketches. Dec. 1910 © SPRI MS:234/4
- Emperor Penguins 1903 © SPRI N:444
p169	- Emperor Penguin chasing chick. c.1905 © private

p170 - White morph Giant Petrel, Cape Adare. 09 Jan. 1902 © SPRI N:1694
- Giant Petrel sketch. 09 Dec. 1910 © SPRI N:1693

p171 - Barograph. *South Polar Times*. May 1902 © DHT

p172 - Field sketch of larger Prion. 06 Mar. 1904 © SPRI Y:2006/15/34
- Field sketch of Albatross sp.. 09 Mar. 1904 © SPRI Y:2006/15/35

p173 - Dines Pressure Tube Anemometer. *South Polar Times*. May 1902 © DHT

p174 - Sledging harness for men. Armitage 1905 © private

p175 - *Prion banksi*, [*Pachyptila desolata banksi*]. Skin No 7. Pack Ice. Nov. 1901 © SPRI N:1668
- Giant Petrel sketch. 19 Dec. 1910 © SPRI N:1693

p176 - Berg seen off Barrier. 02 Feb.1902 © SPRI N:1368

p177 - Emperor Penguin. 1903 © SPRI N:445

p178 - Adélie Penguin sketches. Dec. 1910 © SPRI MS:234/4
- *Oceanites oceanicus*, Wilson's Strorm Petrel. 16 Dec. 1910 © SPRI N:1668

p179 - *Puffinus cinereus* [Grey Petrel] head. 2 Nov. 1901 © SPRI N:1706

p180 - Emperor Penguin tobogganing. Pack Ice. Jan 1902 © SPRI N:1479
- Antarctic Petrel, Pack Ice. Dec. 1910 © SPRI N:1655

p181 - Mt. Erebus from Crater Hill, showing unusual second vent. Undated © SPRI N:1801/64
- Antarctic Petrel, Pack Ice. Dec. 1910 © SPRI N:1655

p182 - Emperor Penguin. 1903 © SPRI N:444
- Three in a sleeping bag, in camp. Undated © SPRI N:21

p183 - *Discovery* leaving Port. *South Polar Times*. June © DHT

p184 - *The Barrier Silence*. Poem by Edward Wilson. *South Polar Times*. Oct. 1911 © SPRI

Endpaper

Blue Whales, from the Crow's Nest.
Pack Ice, Ross Sea. 19 Dec. 1910 7am ©
SPRI N:463

Cover

All cover images are strictly copyright ©
and are acknowledged elsewhere in
this book.
Flaps/bookmarks: all photographs ©
private; Ice Crystals. Undated © SPRI
1803/149

THE BARRIER SILENCE

THE Silence was deep with a breath like sleep
>As our sledge runners slid on the snow,

And the fate-full fall of our fur-clad feet
>Struck mute like a silent blow

On a questioning "hush", as the settling crust
>Shrank shivering over the floe;

And the sledge in its track sent a whisper back
>Which was lost in a white fog-bow.

AND this was the thought that the Silence wrought
>As it scorched and froze us through,

Though secrets hidden are all forbidden
>Till God means man to know,

We might be the men God meant should know
>The heart of the Barrier snow,
>In the heat of the sun, and the glow
>And the glare from the glistening floe,

As it scorched and froze us through and through
>With the bite of the drifting snow.

Ice Island Berg and Pac